W9-DHF-670

MAKING CONNECTIONS

Also in the series:

Making Connections

Technological learning and regional economic change

Edited by
EDWARD J. MALECKI
University of Florida, USA

PÄIVI OINAS
Erasmus University Rotterdam, The Netherlands

Ashgate

Aldershot • Brookfield USA • Singapore • Sydney

Published by
Ashgate Publishing Ltd
Gower House
Croft Road
Aldershot
Hants GU11 3HR
England

HD
58.82
.M34
1999

Ashgate Publishing Company
Old Post Road
Brookfield
Vermont 05036
USA

British Library Cataloguing in Publication Data
Making connections : technological learning and regional
    economic change. - (The organisation of industrial space)
    1. Economic development - Technological innovations
    2. Technology transfer 3. Cooperation
    I. Malecki, Edward J. II. Oinas Päivi
    338.9

Library of Congress Cataloging-in-Publication Data
Making connections : technological learning and regional economic
    change / edited by Edward J. Malecki, Päivi Oinas.
        p.   cm.
    ISBN 1-84014-550-1
    1. Organizational learning. 2.Business enterprises-
    -Communication systems. 3. Intellectual cooperation.
    4. Information networks--Economic aspects. I. Malecki, Edward J.,
    1949-   . II. Oinas, Päivi.
    HD58.82.M34   1998
    338.6--dc21   98-45773
                                                                                    CIP

ISBN 1 84014 550 1

Printed in Great Britain by Galliards, Great Yarmouth

Contents

Part II Becoming Connected

List of Figures

List of Tables

List of Contributors

Neil Alderman, Centre for Urban and Regional Development Studies, University of Newcastle-upon-Tyne, United Kingdom.

Bjørn T. Asheim, Department of Sociology and Human Geography, University of Oslo, Norway.

Györgyi Barta, Centre for Regional Studies, Hungarian Academy of Sciences, Budapest, Hungary.

Philip Cooke, Centre for Advanced Studies in the Social Sciences, University of Wales Cardiff, United Kingdom.

Olivier Crevoisier, Institut de Recherches Économiques et Régionales, University of Neuchâtel, Switzerland.

Martina Fromhold-Eisebith, Department of Geography, University of Technology Aachen, Germany.

Roger Hayter, Department of Geography, Simon Fraser University, Burnaby, British Columbia, Canada.

Edward J. Malecki, Department of Geography, University of Florida, USA.

Peter Maskell, Danish Research Unit on Industrial Dynamics, Copenhagen Business School, Denmark.

Päivi Oinas, Economic Geography/Applied Economics, Erasmus University Rotterdam, The Netherlands.

Sam Ock Park, Department of Geography, Seoul National University, Korea.

Jerry Patchell, Division of Social Science, University of Science and Technology, Hong Kong.

Kevin Rees, Department of Geography, Simon Fraser University, Burnaby, British Columbia, Canada.

Jici Wang, Department of Geography, Peking University, Beijing, China.

List of Abbreviations

ASRDLF	Association de Science Régionale de Langue Française
BEZ	Beijing Experimental Zone
BO	Branch office
CAS	Chinese Academy of Sciences
CIM	Computer-integrated manufacturing
CMTO	Customise-to-order
COMECON	Council for Mutual Economic Assistance
CSIR	Council for Scientific and Industrial Research (India)
CURDS	Centre for Urban and Regional Development Studies, University of Newcastle-upon-Tyne, UK
DA	Design approved
DS	Design supply
EC	European Community
EFTA	European Free Trade Association
ERVET	Ente Regionale per la Valorizzazione Economia del Territorio – the regional development agency (Italy)
ETDZ	Economic and Technological Development Zone
ETO	Engineer-to-order
EU	European Union
FDI	Foreign direct investment
FFPS	Factory, firm, and production system
GATT	General Agreement on Tariffs and Trade
GDP	Gross domestic product
GDR	German Democratic Republic (or East Germany)
GEI	Greatwall Enterprise Institute
GREMI	Groupe de Recherche Européen sur les Milieux Innovateurs (European Research Group on Innovative Milieus)
HIDZ	High- and New-tech Industrial Development Zone

HQ	Headquarters
HTP	Hardware Technology Park
ICT	Information, computing and telecommunications
ID	Industrial district
IGU	International Geographical Union
ILS	Interaction and learning site
ISDN	Integrated services digital network
IT	Information technology
LF	Large firm
MBA	Master of Business Administration
MNC	Multinational company or corporation
MP	Main plant
MSF	Medium-size firm
NICs	Newly industrializing countries
NIS	National innovation system
NRW	North-Rhine Westphalia
OECD	Organisation for Economic Co-operation and Development
OEM	Original equipment manufacturer
PHARE	Poland, Hungary: Aid for the Reconstruction of the Economy
PSE	Public sector enterprise
R&D	Research and development
RIS	Regional innovation system
SF	Small firm
SIS	Spatial innovation system
SMEs	Small and medium-size enterprises
SMIL	Foundation for Small Business Development (Sweden)
STP	Software technology park
TNC	Transnational corporation
UK	United Kingdom
UNIDO	United Nations Industrial Development Organisation
USA	United States of America
WTO	World Trade Organisation

Making Connections: Introduction

Edward J. Malecki and Päivi Oinas

Few phenomena have become as critical to economic activities as the relationships that economic actors form with others in their environments. This is especially the case in technological learning where steady flows of information are crucial for the creation of new ideas. The relationships that firms form with other firms and with other sources of knowledge enable them to deal with changes in technologies, markets, and other aspects of the business environment. The knowledge on which new technologies (and often new industries) is based is both local and nonlocal – even 'global'. Local learning demands that firms be closely linked with, and deeply embedded in, their local economic and social environment where relevant business-specific knowledge is readily available. The experience of industrial districts has suggested that local relations provide key resources that are not likely to be obtainable elsewhere. The local environment also endows a firm with the benefit of institutional support, based on social, economic and political institutions that grow up and evolve suited to the communities – including their firms and industries – in each place. Such an environment provides skills and experience that cannot be easily or quickly duplicated in other places.

Regional economic change can be volatile and subject to sudden shifts in demand and in technological underpinnings, however. Only if firms are attuned to new technological developments and related customer needs will they be able to adapt to pressures for new products, new materials, and new demands regarding service or delivery. To remain competitive, therefore, requires that firms learn new product and process technologies and incorporate these into their established modes of operation. While local relationships are important, it needs to be increasingly emphasized that the creation and maintenance of non-local connections also plays a significant role in sustaining competitiveness as it allows for the incorporation of new ideas into the local processes of technological learning. Especially in high-technology sectors, international networks are required to provide specialized technical knowledge

1

that is still tacit and not widespread. The relative importance of local vs. nonlocal relations, however, remains unsettled in recent research. We need more thorough understanding, in particular, of the nature and extent of the kinds of connections that firms need to establish in the process of sustaining competitiveness in their particular industries.

Contents of the Volume

The objective of this volume is to provide greater understanding of the crucial relationships that are needed between actors in processes of learning at various levels of analysis ranging from interpersonal relations to firms and interfirm networks. The contributions to the volume employ an interdisciplinary set of approaches to technological change (geography, economics, sociology, management studies) to elaborate on these relationships in both intra- and interregional contexts. These processes take place differently in core regions, including Europe, North America, and Japan, in comparison to peripheral regions, including much of Asia, Eastern Europe, Latin America and Africa. The book provides empirical analyses from both core and peripheral regions, highlighting various aspects of 'making connections' under different societal conditions.

The papers gathered here were initially presented at a conference of the IGU Commission on the Organisation of Industrial Space in Göteborg, Sweden, in August, 1997 and subsequently revised substantially. They build on the knowledge gained in previous research of the Commission and published in earlier volumes in the Organisation of Industrial Space series, and expand our knowledge in two directions.

The first part, *Connections and Competitiveness*, explores the basis of innovation and of learning as the fundamental ways in which firms adapt to new conditions. We begin in the first chapter by outlining a view on 'spatial innovation systems' which captures the idea of the importance of *both* national and regional innovation systems but also emphasises the importance of connections between various regional innovation systems in technological development. Peter Maskell provides a compelling reason for the necessity of connections – that all new knowledge becomes not only old but also imitated and, eventually, ubiquitous and no longer a source of competitive advantage. He stresses the importance of local connectedness and the resultant opportunities for localised learning, but his analysis clearly supports the need for external connections as well. Olivier Crevoiser outlines what is special about cities for the innovation process: that they provide interaction and learning sites that facilitate contacts, learning, and commercialisation of technology.

The authors of the following two chapters probe the challenge that all firms face to build networks of contacts both within their local area and with critical sources often in more distant locations. Neil Alderman finds in his empirical research in Britain that engineering firms do not seem to rely on local networks, preferring instead to develop close links with key customers, including those in foreign locations. Jerry Patchell, Roger Hayter and Kevin Rees examine a particular group of firms that grow to large size and whose growth depends critically on being innovative within a specialised product-market niche. They rely upon close links with customers and suppliers to create and sustain state-of-the-art products and capabilities – and often market leadership.

The second part, *Becoming Connected*, deals with the experience of regions in various countries that are attempting to build and to expand their networks of contacts and information. Bjørn Asheim and Phil Cooke provide a theoretical discussion on the characteristics of localised interactive innovation networks and the regional innovation systems that they give rise to. They present a large number of empirical examples of European regions where variants of learning and interactive innovation networks have developed. Györgyi Barta describes the situation in Hungary, where the post-socialist transition has rendered old connections useless and many new connections had to be made to the capitalist marketplace. The new connections, made largely through multinational corporations, have brought techniques and management skills, but at the price of becoming dependent on these powerful organisations. Jici Wang reports on an aspect of the Chinese transition to a socialist market economy, namely new entrepreneurship in high-tech industries such as electronics. She discusses the problem that little interfirm networking has taken place among high-technology firms despite their concentration in a single zone within the Beijing municipality. She views this phenomenon mainly as a legacy of the hierarchical bureaucracies of central planning. Martina Fromhold-Eisebith then provides the example of Bangalore, India, where dense social and interfirm networks create the basis for the flourishing of a cluster of IT firms, both Indian and branches of MNCs. Finally, together with Sam Park, we look at the pivotal mechanisms for technology-based development and dwell upon regional and national policies to support such development.

The relationships that firms form with other firms and other regional actors form with other firms and different sources of knowledge enable them to develop competitive capabilities amid changing technologies and markets.

This competitiveness directly affects the economic development of places and the lives of people living in them. The relative importance of local vs. nonlocal relations is still unsettled in recent research, and this volume aims to take some steps towards shedding light on the importance of both.

PART I
CONNECTIONS AND COMPETITIVENESS

1 Spatial Innovation Systems

Päivi Oinas and Edward J. Malecki

Introduction

It is generally acknowledged that technological development is the driving force behind social and economic development. However, there are still many open questions concerning how technological development takes place – and how it takes place in space. Technology is in many respects 'everywhere'. It affects what work we do, how we work, where we work, when we work, and with whom we work. And it affects us while not working: leisure activities are increasingly 'personalised' through technology – personal computers, personal concerts on WalkMans, and personal information downloads from the Internet.

Despite the (sometimes world-) wide diffusion of many results of the advancement of technology in terms of products used, its developmental trajectories follow a more limited set of paths. Moreover, these trajectories are controlled by a limited set of actors in time and space. The evolution of technology and the industrial structures that accompany technologies have certain life-cycle characteristics (Nelson, 1996). These cycles include stages during which products, processes, and ideas emerge and standards and dominant designs evolve. In most industries, unexpected technological discontinuities also intervene to disrupt established patterns and dominant firms (Tushman and Rosenkopf, 1992; Utterback, 1994). These life-cycle patterns, in turn, have spatial patterns.

This chapter aims at examining existing literature on technological change in a manner that allows us to understand how technological change takes place in processes that are tightly spatially bound and as a whole can be seen as forming *spatial innovation systems* (SIS). We regard the creation of innovations – as well as their adaptation and adoption (and non-adoption) – as inseparable from the local/regional socio-economic circumstances in which they take place. In contrast to approaches aiming at identifying 'national systems of innovation', we take as our starting point the view, ingrained in recent thinking, that success in technological competition is based on the ability of specific *sub*-national and *trans*-national technological systems to produce flows of innovation and to keep up with state-of-the-art practices in different technological frontiers, as well as their societal application.

A distinction is made between the systems that are capable of producing innovations and those that facilitate their adaption and adoption. Both of these systems have distinct spatial dimensions, the identification of which helps us understand how technological change takes place in various circumstances.

Spatial Innovation Systems: Potentially Multiregional and Multinational

Since Freeman's (1987) introduction of the concept of *national systems of innovation*, his approach has gained widespread recognition within the literature on technological development and change. However, an increasing awareness has grown among those sensitive to spatial issues about the fact that it might make equal sense to talk about *regional innovation systems* (RIS; see, e.g., Asheim and Cooke, Chapter 6, *this volume*). This awareness does not claim that the national scale as the context for innovative activity (including factors such as national government policies, state supply of venture capital, other national actors, national cultures, and industrial practices) would not play an important role in technological development. However, within nation-states, specific regions tend to bring about a large share of the outcomes that are regarded as the accomplishments of national systems of innovation in the respective literature.

The economists Archibugi and Michie express a view that appears to be in line with how economic geographers approach technological development:

> To understand technological change it is crucial to identify the economic, social, political and geographical context in which innovation is generated and disseminated. This space may be local, national or global. Or, more likely, it will involve a complex and evolving integration at different levels of local, national and global forces (Archibugi and Michie, 1997, p. 122).

Indeed, it is the nature and details related to the 'complex and evolving integration at different levels' that have to be explicated in research interested in understanding how different systems of innovation operate. It is not clear from Archibugi and Michie's discussion, however, what kind of role a subnational (local) scale would play in the developments that they, later on in the text, end up discussing either at the national scale or 'globally'. As a matter of fact, as Archibugi and Michie continue their discussion, it seems that the subnational scale ('local') is 'subordinate' to the national one.

The scale that we refer to as local or regional in this chapter stands on its own and may be more significant in some countries than the national scale as far as technological development and innovativeness is concerned. Archibugi and Michie (1997, p. 128) point out that 'we are still far from having achieved a coherent conceptual and empirical framework with which to explain the diversity between different countries' success in innovating'. Part of this inability may well stem from the inability to account for interregional differences in innovativeness within nations. It needs to be acknowledged, that regional disparities in innovative capabilities vary significantly among countries.

Archibugi and Michie (1997, pp. 127-128) elaborate on issues such as education and training, science and technology capabilities, industrial structure, S&T strengths and weaknesses, interactions within the innovation system, absorption from abroad. While there certainly are country-specific differences in these aspects as they depict them, it is quite clear that the same applies to regions as well. Regions within countries share some of the aspects of the entire nation (such as educational legislation and technology policy) but they have different possibilities to 'go their own ways', and ultimately end up diverging from a 'national average'. Small industrialised countries might experience interregional discrepancies to a lesser degree than larger and/or developing countries but as regional actors try to create competitive advantages, they actually try to distinguish themselves from others (e.g., Malmberg and Maskell, 1997a, 1997b; Maskell, Chapter 2, *this volume*; Maskell and Malmberg, 1998; Ohmae, 1995).

Especially if there is a larger number of firms that form a specialised local system, it is likely that the region becomes distinct from the rest of the regions in a country in some significant respects. It has been suggested that the most 'innovative regions' are those that host agglomerations of single or highly related industries (OECD, 1996; Malmberg and Maskell, 1997a). However, successful – and especially innovative – systems may also arise from *combinations of industries* in specific locations, if firms operating in those industries build linkages between them (Oinas and Van Gils, 1998). Harrison *et al.* (1996) believe that diverse locales (i.e., locales with relatively large numbers of industries) are more important for promoting innovative firm behaviour than localized clusters of firms in a single industry.

Since the flexible specialisation and localisation enthusiasm of the 1980s, it has become clear that regional (or local) systems are not 'self-contained' entities but are linked to 'the outside world' by various sorts of connections (Amin and Thrift, 1992, p. 572). These connections help actors within the regional systems to stay tuned with what happens in the market, what happens among other producers (both competitors and collaborators), among consumers, among scientists, among regulators, support agencies and other sources of technological knowledge. Indeed, it seems increasingly clear that the connections of regional actors to extraregional ones stand as momentous in gaining competitive advantage through technological progression.

Thus, in our discussion, what we call *spatial innovation systems* refer to *regional innovation systems* plus *their interconnections*. The regional system links global knowledge of generic technologies to specific applications produced by local firms (Arcangeli, 1993) which combine existing technologies (using both generic and tacit knowledge) in new ways to bring about genuinely new innovations. In consequence, spatial innovation systems consists of overlapping and interlinked national, regional and sectoral systems of innovation which all are manifested in different configurations in space. We shall outline an approach to the spatial innovation systems in the subsequent sections below, characterising the different kinds of places (regions or localities) that are involved and the importance of interrelations between those places. In these systems, some regions are – to significantly greater extents than others – 'producers' of innovations, or 'innovators'. For those regions, internal and external relations play a significant role. Other regions can be characterised as 'adapters' and 'adopters'. The truly peripheral regions, which remain largely outside these systems, may be thought of as 'non-adopters'.

Innovators: Best Practice via Distinctive Regional Relations

Regional Networks

A great deal of research effort has – justifiably – gone into understanding the state-of-the-art regions: those places that qualify as 'learning regions', as 'innovative milieus', as 'clusters', as 'industrial districts', or as 'local productive systems'. Cooke (1996), for example, discusses a number of places that have achieved economic excellence and competitive advantage, such as Baden-Württemberg, and Sternberg (1996) has evaluated the factors behind several successful regions. Although these studies provide clues to understanding regional development, they can only brush the surface of complex processes that are deeply entwined in the regional characteristics.

These characteristics include both what have traditionally been called *economic* (dubbed here the 'hard' aspects of economic action: the measurable items that are traditionally dealt with in economic analyses) and *socio-cultural* (the items that are difficult to conceptualise and often impossible to measure: the 'soft' aspects of economic action).

It is often suggested that the ideal institutional environments are to be found in what are called 'industrial districts' or 'innovative milieus'. Indeed, much of our understanding of the crucial interfirm networks comes from research into such areas. Industrial districts and milieus embody the interaction and dense network of linkages that comprise a local production system, usually around a single or highly related industries (OECD, 1996). Because of these linkages, the territory itself is said to function as a collective entrepreneur with firms, as well as inter-firm associations, worker organizations, financial institutions, and governmental agencies playing supporting roles (Best, 1990; Morgan, 1996).

What the GREMI research group calls 'innovative milieus' embody the necessary interactions in a particular way. Such a locality is comprised of a set of relations which unite a local production system, a set of actors and their representations, and an industrial culture (Camagni, 1995, p. 195). In an innovative milieu, two sets of effects operate simultaneously: *proximity effects*, such as reductions in costs because of quicker circulation of information, face-to-face contacts, and lower costs of collecting information; and *socialization effects*, related to collective learning, cooperation, and socialization of risks. These two processes are *collective rather than explicitly cooperative* and thus spread beyond any single bilateral interfirm linkage. Dynamic processes of interaction and of learning characterise such milieus (Becattini, 1991; Maillat, 1995, 1996).

Asheim (1996), for example, criticizes the innovative milieu approach for failing to 'specify the mechanisms and processes which promote innovative activity more successfully in some regions than in others'. Even though these processes are not fully identified in the literature on industrial districts, innovative milieus and other successful areas, three features stand out as important. First, the collectivity that somehow encompasses regions in their entirety, second, the emphasis they put on the 'soft' aspects of economic activity, and third, the recurrent theme of extralocal relations (yet, the last issue is mainly paid lip service rather than fully discussed).

Several research findings point to Silicon Valley as a unique environment for technologically advanced entrepreneurship. Silicon Valley provides a large number of entrepreneurial role models and examples. This environment comprises an unusual 'ecosystem' within which firms form and re-form through continual entrepreneurship. Networks of interpersonal relationships support entrepreneurship, links among enterprises (large and small alike), and innovative activity (Bahrami and Evans, 1995; Saxenian, 1994). The networks provide the 'soft' bonds that link the 'hard' qualities of Silicon Valley, which take the form primarily of high levels of state-of-the-art R&D. The hard traits also include several of the characteristics of business clusters: proximity of suppliers, capital availability, access to specialized services, and machine and tool builders (Rosenfeld, 1997). Imitation of the 'hardware' of Silicon Valley, such as through creation of a science park, has not been effective in most places. As Massey *et al.* (1992) and Johannisson *et al.* (1994) have shown, interaction does not necessarily take place despite geographical proximity; something else is needed. This 'something else' often is described simply as 'synergy' (Castells and Hall, 1994; Stöhr, 1986), but presumably constitutes the 'software' of a region: the presence of social structures of sociability, trust, and an industrial structure that demands interaction among firms (e.g. highly linked industries making flexibly changing products) (Amin, 1994). This synergy typically fails to develop in 'created' – rather than spontaneous – research parks (Luger and Goldstein, 1990; Quintas *et al.*, 1992).

Unlike Silicon Valley, where high-tech entrepreneurship is the dominant process, Germany's Baden-Württemberg derives its strength from the region's institutions. The renowned system in Baden-Württemberg provides training, information and R&D that support innovation. Small firms do not have to bear the burden of flexible production; 'many of the costs ... are shared by or embedded in a deep network of organizations and practices in the political economy' of the region (Herrigel, 1993, p. 17). Baden-Württemberg is regarded by many as a model of a networked region, in contrast to the *dirigisme* represented by French or Japanese technopoles or the grassroots *Kohsetsushi* centers (Cooke and Morgan, 1994a). In some ways firms have been discouraged from collaborating, however, since they have been able to draw upon local support institutions and upon large firms such as Robert Bosch, rather than upon other small firms (Grotz and Braun, 1993; Mueller and Loveridge, 1995). Increasing globalization and the necessity for the skill-intensive R&D of flexible production have caused a serious need for collaboration so that the regional system of innovation does not break down (Cooke and Morgan, 1994a, 1994b).

Rosenfeld (1997), who believes that clusters of interdependent firms are the enabling phenomenon for networks, lists several factors that influence the 'power of clusters to produce synergy'. Many of these are 'soft' factors (such as entrepreneurial energy, innovation and links to innovative networks elsewhere, and shared vision and leadership) which amplify the 'hard' factors associated with the local economy. The degree of local interaction of firms with their environment (including various sorts of support institutions and other firms) is critical. Firms must cooperate, share information and resources, and tackle problems jointly (Rosenfeld, 1997, p. 15). In ideal settings, firms are supported by strong local government and non-governmental institutions and the provision of a wide range of social services (Perrow, 1992). Interaction that takes place in trust-based relationships provides information that can be put to use in enterprises and results in successful market and non-market transactions.

In the following we will leave behind the discussion of *types of regions* ('industrial districts', 'innovative milieus', 'clusters'), and *success stories* (such as Silicon Valley or Baden-Württemberg). We will discuss four concepts that have become recurrent in recent literature on technological development in the regional context, namely *learning, embeddedness, institutional thickness* and *conventions,* and consider what they teach us about technological development in space.

The usage of these concepts presupposes that interfirm networks are essential for technological development at the local and regional scale. Where such interaction is not present, local technological development suffers (Bull *et al.*, 1991; Hardill *et al.*, 1995). Not all spatial agglomerations of firms in the same or related sectors necessarily comprise an industrial district or local industrial system in a true sense. Furthermore, to refer to network formation as such is too 'neutral' and does not lead very far in explaining what is the crucial thing that happens in network relations. Namely, from the point of view of competitive success, what is important in networks is that 'right' things get to be done in the competitive markets of each producer. It is increasingly assumed that the 'right' things are found when actors are tied by their networks to particular places. Much of the current wisdom concerning what else is needed – and needs to be better understood – is elaborated in the following discussion.

Learning and the Regional Culture of Learning

Maillat (1995) observes that two distinct processes must be operating for an innovative milieu: *interaction* and *learning.* Something less than this results from either an emphasis on interaction in the absence of learning (e.g. industrial districts, with the threat of lock-in), or on learning in the absence of interaction (e.g. technopoles, isolated from their regional environments). Indeed, along with a vast and growing interdisciplinary literature concerning learning in organisations and in entire economies, learning has become the key (fashion) word in regional and technology development literature in the 1990s as well. Current thinking suggests that the technological vitality of regions revolves around their learning efficiency. Accordingly, evolving into what some authors call a 'learning region' (Asheim, 1996; Florida, 1995a, 1995b; Maillat, 1995; Simmie, 1997) appears to be widely regarded as the 'best practice' for any region. Learning is viewed as the ultimate virtue, because it represents the capability to respond to new situations, new opportunities, and to participate in the process of creating emergent new technologies.

Because no firm can do everything itself, external relations are essential; learning also often occurs through these relations (Eliasson, 1996), and can be facilitated by the regional environment (cf. Asheim and Cooke, Chapter 6, *this volume*; Maskell, Chapter 2, *this volume*). The creation of a skilled workforce, for example, falls largely on the region. Regional industrial knowledge and skills are created and transmitted to future generations through education, training, and acquired experience. Essential industrial skills, however, frequently are not taught in schools; often tacit knowledge and skills are passed down by practical experience. The learning approach does not make a predetermined division into 'innovative' and 'non-innovative' activities, sectors or regions – such as those related to research and development or high technology (Oinas and Virkkala, 1997, p. 265). Instead, knowledge creation is regarded as an integral part of all sorts of activities; processes of learning are associated with all stages of production (Maillat and Vasserot, 1988; Asheim and Cooke, Chapter 6, *this volume*). Learning takes place in various phases in technological change as problems are solved, in carrying out old practices, becoming aware of problems, developing new solutions for existing problems and redefining problems and implementing new ideas (cf. Schienstock, 1996). It is also usually recognised that the actors that are involved in processes of learning include both various types of firms and public or quasi-public actors. Yet, actors in some places are more open to new ideas. Regions where imports are state-of-the-art and diverse are more likely to be innovative themselves.

The learning approach emphasises the collectivity of problem solving; '[t]eams of R&D scientists, engineers and factory workers [are] collective agents of innovation' (Florida, 1995b, pp. 528-529). The collective aspect of learning sometimes comes up somewhat naively in the enthusiast usage of the 'learning region' metaphor: as if everyone involved in them aimed at the joint goal of creating competitive advantages (cf. Oinas and Virkkala, 1997, p. 270); as if 'learning regions' were happy collectively learning communities where no sign of friction nor domination is to be found – too heavenly to be descriptions from the earth. The expression 'regional learning' appears to be a more appropriate term (Oinas and Virkkala, 1997, p. 270). In practice, who learns and what will depend on the opportunities available to each individual participant, and on what is regarded as relevant 'search' areas in processes of learning. Both tend to be influenced by powerful actors involved in learning processes. In order for learning to take place often requires that actors can scan for information. Most small firms, e.g., cannot (Johannessen *et al.*, 1997). Within firms, the ability to scan increases with age and size, most likely because knowledge and the capability to add to existing knowledge are cumulative (Mohan-Neill, 1995). Under favourable conditions, however, even smaller firms may be able to access new information and the knowledge needed for its use, e.g., through regional interfirm and social networks.

The Embeddedness of Technological Innovation in Structures of Social Relations

To an extent, the 'systemness' (Tödtling and Sedlacek, 1997) of many successful regions is captured by adopting the Polanyian-Granovetterian concept of embeddedness (Granovetter, 1985; Amin and Thrift, 1994a, 1994b). By now, it is accepted that 'embeddedness' significantly influences how economic actors operate. The problem with this wisdom is that 'embeddedness' tends to be used without specifying what it means (Oinas, 1997). The plainest (perhaps the most common) usage refers to structures of network relations of economic actors (e.g., Uzzi 1996, 1997). From the geographical point of view, what has been stressed most are *local linkages* (Crewe, 1996; Dicken *et al.*, 1994; Turok, 1993). Economic sociologists tend to emphasise Granovetter's (1985) discussion on trust as the basis of social relations (e.g., Uzzi, 1996). The issue of trust is beginning to enter geographers' discourses as well (e.g., Harrison, 1992) even though there is not much in-depth research into it. Trust is supposed to provide economies of time in information gathering (an important advantage for small firms especially). Uzzi (1997) has shown that a firm – indeed, an entire network – can become overembedded, if too few ties exist, or if they are not sufficiently diverse to respond to external forces. Grabher (1993) has pointed out that this can be a weakness of regions as well.

These considerations, however, seem to be just part of what Granovetter (1985) hinted at in his seminal research agenda. In our discussion, because competitive advantage via innovative activity tends to be built on the basis of unique resources (Maskell, Chapter 2, *this volume*), innovative environments must be agglomerations or localized production systems where several conditions come together to provide advantages that are not available everywhere, and perhaps nowhere else (Lipietz, 1993; Maskell and Malmberg, 1998; Porter and Wayland, 1995; Storper, 1997). If the embeddedness of actors in social relations in such locations is to be regarded as significant for technological innovation we need to understand what 'happens' in embeddedness; what is the 'content' of embeddedness. Recent accounts suggest that the embeddedness of economic action in social relations involves also – or maybe especially – the cognitive, cultural and political influences on economic action (Zukin and DiMaggio, 1990; cf. Dicken and Thrift, 1992). The need to understand these issues stretches from descriptions of network relations at the local scale to those that cross national boundaries and involves the broad variety of cultural and political influences that are transferred in such embedded relations (Oinas, 1997). These potentially very different kinds of relations need to be 'managed' in specific places and regions.

Social Relations Characterised by Institutional Thickness

Amin and Thrift (1994a) have launched the notion of institutional thickness to characterise favourable conditions in regions that 'succeed' in global competition. At the outset, thinking of how 'institutional thickness' and 'embeddedness' are used in the literature, they seem to refer to a highly related – if not the same – phenomenon. It is possible to regard the notion of *institutional thickness* as referring to *specific characteristics of the social relations (in total) in which actors are embedded*, while *embeddedness* refers to the *nature of the relation to the total set of relations* from individual actors' points of view. (*Embedding*, then, refers to the process that takes place as actors become embedded or act as embedded). Discussing institutional thickness and embeddedness from this perspective gives them distinct meanings.

For Amin and Thrift, institutional thickness is the kind of 'liminal concept' that holds 'the key to understanding the workings of the global economy' (1995, pp. 101-102). They characterise institutional thickness by isolating four factors from the literature on institutions and economic development (Amin and Thrift, 1995, p. 102; cf. Amin and Thrift, 1994a, pp. 14-15):

•   a strong presence of a variety sorts of organisations[1] '(including firms, financial institutions, local chambers of commerce, training agencies, trade associations, local authorities, development agencies, innovation centers, clerical bodies, unions, government agencies, ... business service organisations, marketing boards, and so on), all or some of which can provide a basis for the growth of particular local practices and collective representations in social networks';

- high levels of interaction amongst the various organisations in a local area;
- 'sharply defined structures of domination and/or patterns of coalition resulting in both the collective representation of what are normally sectional and individual interests, and the socialisation of costs and the control of rogue behaviour';
- 'a mutual awareness that they are involved in a common enterprise'.

At the core of 'institutional thickness' seems to be *(established) local practices and collective representations in social networks* (mentioned as part of point 1) which could be interpreted and elaborated as the glue or adhesive that holds the other characteristics together. This, through the high levels of interaction among local actors (point 2) give rise to locality-specific institutionalised control on the one hand (point 3) but also to the feeling of belonging to a community aiming at a common goal (point 4) on the other. The control aspect (point 3) points at the need to recognise the powerful actors among the many others involved, and their role in the creation of the circumstances of 'thickness' (firms, individuals, support organisations etc.). In Amin and Thrift's account (1994a, p. 15), institutional thickness will, 'in the most favourable cases', produce the following outcomes: institutional persistence; the construction and deepening of an archive of commonly held knowledge of both the formal and tacit kinds; institutional flexibility, which is the ability of organizations in a region to both learn and change; high innovative capacity; the ability to extend trust and reciprocity; and the consolidation of a sense of inclusiveness, that is, a widely held common project which serves to mobilize the region with speed and efficiency. These are of interest for the search to understand the local or regional bases of spatial innovation systems.

We believe that institutional thickness (and the ensuing conception of regions as kinds of collective entrepreneurs, through the 'widely held common project') is likely to be an appropriate characterisation only for the dynamic parts of a few densely networked and specialised industrial localities consisting of actors engaged in innovative activities. There is, however, a need to unravel the qualities of such successful areas whose success goes beyond simply individual actors or their public and private networks to function as a collective asset for the actors involved. 'Institutional thickness' is an interesting hypothesis in this regard.

Social Relations Coordinated by Conventions

If, as suggested above, established local practices and collective representations in social networks are at the core of the qualities of successful regions dubbed 'institutional thickness', Storper and Salais' recent writings seem to hit at that very same core. In addition to the notion of *untraded interdependencies* (Dosi, 1988), the technological externalities that become a collective regional asset for the firms involved (Storper, 1993, 1995, 1997), Storper and Salais (1992, 1997; Storper, 1993, 1997) claim to have found the more fundamental notion, that of *conventions*.

> Much has been written recently about the role of 'non-economic' forces, such as institutions, cultures, and social practices, in economic life. We intend to take those forces as central to the economic process and no longer consider them as 'non-economic'. They take the form of conventions (Storper and Salais, 1997, p. v).

What this idea conveys for our discussion is that the 'rationality' of groups of innovators can be seen as formed by conventions which 'include taken-for-granted mutually coherent expectations, routines, and practices, which are sometimes manifested as formal institutions and rules but often not' (Storper, 1997, p. 38). Thus, conventions are the carriers of the pragmatic knowledge, ways of thinking and believing related to technological innovation, possibly also about its future directions. The usage of 'conventions' highlights the importance of collectively taken-for-granted practices for innovation. 'When actors undertake an activity, they do so with the expectation that they have a framework of action in common with other actors engaged in that activity' (Storper, 1997, p. 45). Thus, conventions bring about *coordination* (Storper, 1997, pp. 42-43) into innovative activities. They facilitate interactive processes and the needed mutual understanding in the exchange of ideas (Storper, 1997, p. 45) in the various delicate stages of producing innovations, and they help reach solutions concerning the 'right things' to be done to generate economically viable innovations (Storper, 1997, p. 126; Storper and Salais, 1997, p. 4). Conventions 'have cognitive, informational, and psychological and cultural foundations' (Storper, 1997, p. 43). Conventions are manifested in variations in regional learning efficiency. These are frequently attributed simply to cultural norms that are not easily transplanted or created (Saxenian, 1994; Sweeney, 1996).

The problem with the concept of conventions is that it takes so many forms; it covers such a vast array of qualities of actors' beliefs, behaviours, external relations, and material conditions (see, e.g., Storper, 1997, p. 136) that, *as such*, conventions do not explain a great deal. Yet, the concept does invite further theoretical elaboration as well as detailed empirical investigation, especially into the non-material ('soft') aspects of economic activity that are increasingly believed to have a significant effect on economic outcomes.

Concerning the importance of the local relations of innovative actors, it can be summarised that it is a question for further empirical and theoretical elaboration to delve into the specific *manifestations and outcomes of the embeddedness of innovatiove actors in local networks of social relations characterised by institutional thickness and coordinated by often locally-specific and taken-for-granted conventions*.

Examples of local conventions are found in the unmeasurable connections related to family ties, cultural closeness, and specialized knowledge identified in the industrial districts in the Third Italy. These interfirm relationships are seen to go beyond material flows alone. The success of Silicon Valley is summarized around the point 'that culture, irritatingly vague though it may sound, is more important to Silicon Valley's success than economic or technological factors' (Micklethwait, 1997, p. 7). Micklethwait concludes, following Saxenian (1994), that the following characteristics – perhaps best described as conventions – are present to a greater degree there than perhaps anywhere else in the world:

- tolerance of failure;
- tolerance of treachery;
- risk-seeking;
- reinvestment in the community;
- enthusiasm for change and rapid response to technological change;
- promotion on merit and openness to immigrants and women;
- obsession with the product and the state-of-the-art 'cool idea' in technology;
- collaboration;
- variety of firms in size and specialization;
- easy entry.

Thus, without claiming that the collective learning about regional economies would have reached the level of detailed explanations concerning the nature and importance of local or regional relations for innovation, we seem to have at least some clues of innovation being tightly related to a broad array of factors ranging from the creation and usage of tacit and codified professional knowledge, established modes and patterns of interaction within networks to long-standing, regionally institutionalised behavioural patterns.

External Relations

The above discussion elaborates on what can be said to be key factors in the constitution of innovation systems at the regional scale (RIS). In the following, we will add the insight of the idea of spatial innovation systems (SIS), which emphasises the importance of occasional or sustained interrelations and interaction between the various RISs – keeping in mind that all of those are grounded in their specific national innovation systems (NIS).

The importance of links to nonregional networks is a recurrent finding in recent research on industrial districts and technology districts (Amin and Thrift, 1992; Maillat, 1995; Mueller and Loveridge, 1995; Storper, 1993). Yet, we do not seem to understand the nature and relative significance of external connections very well to date. Connections to other networks in other regions provide access to a diversity of ideas and bases for comparison with local practices that are not internally generated (Amin and Thrift, 1992, 1993; Camagni, 1995; Conti, 1993; Maillat, 1995; Tödtling, 1995). In places where a milieu is not present, however, these external relations in fact may weaken existing local ties (Bull *et al.*, 1991; Curran and Blackburn, 1994; Hardill *et al.*, 1995).

Within regions, each sector has its specific connections to extra-regional partners, which enhance the innovative potential of each sector's actors. In the case of multisector regions, these external relations also add to the total innovative potential of the region's actors in the form of a higher and more diversified technological capability (Oinas and Van Gils, 1998). The reasoning involves a combination of the spatial system of innovation and the technological system which operates across industrial sectors (Carlsson, 1994).

The role of *active, 'extroverted' firms* needs to be acknowledged in their role of making connections (Patchell *et al.*, Chapter 5, *this volume*). Not all firms are outward-looking. Kelley and Brooks (1992) distinguish between firms with primarily active and social external linkages and those with passive and asocial linkages (see also, Amendola and Bruno, 1990; Estimé *et al.*, 1993). 'Passive' firms rely on written media and routine inputs. 'Active' firms, by contrast, utilize written sources as well as personal contacts with sales representatives, participation in trade shows, contact with vendors, and close relationships with special-order customers for sharing of technical information. The presence of external relations of active firms, compared to their absence among passive firms, more than triples the probability of adoption of production automation among the smallest firms (fewer than 20 employees) and more than doubles it among firms with 20-99 employees (Kelley and Brooks, 1992). Thus, if small firms seek out and obtain technical information from external resources, they are much more likely to compensate for their size limitation in the adoption of new technology (Julien, 1995; Rothwell, 1992). The most likely firms to be active in seeking out external information are those with in-house R&D activity (Tsipouri, 1991). Active firms generally exhibit better performance, which results from a combination of in-house R&D, external technical linkages, and internal innovativeness (MacPherson, 1992). Even in peripheral locations, externally oriented firms are able to overcome the constraints related to their location (Alderman, chapter 4, *this volume*; Vaessen and Wever, 1993). Vaessen and Keeble (1995) have found that growth-oriented firms do more R&D and have more external programs for worker training – regardless of their regional environment. Thus, the wider networks of active, extroverted firms tend to encompass more connections both within their own region and outside it. They are also more likely to aim at competition in international markets (MacPherson, 1995).

The growing tendency for some companies to utilize several 'home bases', including R&D and sophisticated production, rather than a single headquarters, for important markets illustrates the need for extraregional connections. This is important to keep in mind as technological capabilities outside the Triad economies are likely to become increasingly important – even if the degree to which R&D activities are 'globalized' and dispersed to more countries is still a matter of dispute (e.g., Archibugi and Michie, 1995; Cantwell, 1995; Cooke, 1997; Dunning, 1994; Patel, 1995).

Even though Silicon Valley is considered the archetypal success region by many, others are skeptical about the entrepreneurial success in Silicon Valley. Harrison (1994) believes that it is less an entrepreneurial industrial district than a hub of global networks of external connections of large corporations. Hobday (1994) suggests that the region's networks function solely for innovation but lack the complementary assets (such as marketing and distribution) needed for later stages of the product life cycle. Yet, it could be questioned why those complementary assets even should be in Silicon Valley. Given Silicon Valley's specialisations it is possible to think of having marketing and distribution handled by firms specialised in those functions and located elsewhere (where, possibly, the local environments support those activities).

Places Below Best Practice

Environments for Adapting Innovations

While the main emphasis and interest of the scientific community seems to have been in the regions that host innovative actors, technological innovation is not absent from less-than-best-practice regions. Regions hosting firms and production systems that are not 'leading edge' or 'best practice' (in the sense that fundamental changes in products and processes are unlikely to originate there), they may fare fairly well in interregional competitive games of investments anyhow. They may do it by providing an environment for steady improvements, incremental innovations, leading to high quality. The ability to learn from innovative firms in other places is considered the best route for developing – and maintaining – innovative capability of this sort (Kim 1997; Mody *et al.*, 1995). We may characterise these places as *adapters of innovations*: not the best innovative environments, but economically competitive and providing good, skilled jobs for their inhabitants.

Lately, research has begun to unravel what is 'wrong' with, or missing from, places that are agglomerations yet seem to lack some key elements. Firms in such agglomerations tend not to be open to outside ideas, are unable to 'unlearn' old ways, and prefer top-down – rather than bottom-up – network structures (Hassink, 1997). What is sometimes most difficult in such adapting environments is to form symmetric network relations with internationally operating actors and information networks, yet this is perhaps the most important task of all (Malecki and Tödtling, 1995). Examples of adapting regions include the NICs of Southeast Asia, where incremental innovations are becoming common (Kim, 1997; Leonard-Barton, 1995; Singh, 1995).

Environments for Adopting Innovations

Where networks – both intraregional and extraregional – are few, innovations are unlikely to originate or to be adapted in meaningful ways. These regions rely on their skilled labor forces to attract outside investment, and struggle not only to catch up but also to make changes. More importantly, Sternberg (1996) generalizes from an analysis of seven well-known high-tech areas (Silicon Valley, Greater Boston, Research Triangle, Western Crescent, Cambridgeshire, Munich, and Kyushu) that the most significant factors behind their success are government R&D expenditures, agglomeration (city size), the research and education infrastructure, and the age of the region. He concludes that the growth and development of high-tech regions is determined by a number of factors similar to those proposed by Porter's (1990) model: interrelated production networks of large and small enterprises, endowments of production factors such as skilled labour and risk capital, the demand for new knowledge-intensive products, and by entrepreneurial strategies and competition. To these he adds a factor more significant at the regional level: 'both implicit and explicit technology policy' (Sternberg, 1996, p. 80).

Bangalore, India (Fromhold-Eisebith, Chapter 9, *this volume*), parts of Mexico, and the Zhong'guancun area of Beijing, China (Wang, Chapter 8, *this volume*) typify this environment. These areas attract a great deal of foreign direct investment, based on their productive workers, but they have not yet attained the perception from the outside as generating either incremental or fundamental innovations.

Peripheries: Late- or Non-Adopters of Innovations

What about the peripheries – where information arrives late, where education levels are lower than elsewhere, and where economic development is a challenge? What Simmie (1997), e.g., examines as 'peripheral regions' include Wales and peripheral Japan, although many observers would not classify these regions as peripheral other than relative to their national cores. The situation is far worse in the poorest regions of the world outside the core regions of the Triad economies (Western Europe, North America, Japan and East Asia).

From the point of view of technological development, there is an obvious scarcity of R&D, especially industrial R&D (Ewers and Wettmann, 1980). And a prominent shortcoming of peripheral regions seems to be their low potential even to adopt innovations. This is an outcome of the obvious reasons related to lack of capital and physical infrastructure, low levels of education, etc. In these settings, contact with the outside is difficult and, therefore, infrequent. There are few opportunities to confront global best practices which prevents technological progress – for good or for worse – from being made. And even in the case of existing (typically governmental) R&D facilities, there is no guarantee of spin-offs, even in the long run. Some places with considerable R&D have not been able to generate new firms in significant numbers, lacking synergy and local networks and competitive clusters of firms (Castells and Hall, 1994). While it is possible to recreate the 'hardware' of Silicon Valley through science parks, it is more difficult to create their 'software', the synergy that embodies the sociability, trust, and industrial structure that fosters interaction among firms. The failure to create the hardware in many advanced places (Sternberg, 1996) implies that the difficulties are even greater in less advanced, and more peripheral, places.

Reaching for Best Practice

The extreme peripheral regions are characterized by 'introverted' firms and/or branch plants and little in the way of learning as an everyday activity. Policies for 'in-between' places, can be tried in an attempt to stimulate interaction and learning, but much depends on just how peripheral a place is. A location in the periphery of the Netherlands is rather different – and far less 'out of the loop' – than a place in a remote region of Australia, Canada, or the USA. Telecommunications technologies can substitute to some degree for remoteness, but active engagement in personal interaction locally or non-locally are key to much of what makes a firm and its regional environment successful.

The shortcomings relate to peripherality – and often to firm size. As an indication of the difficulty of 'being connected', Rosenfeld (1992, p. 130) describes the problem that small rural manufacturers in the USA face in 'obtaining objective information, absorbing it, and evaluating the knowledge gained'. This lack of awareness is not ignorance, but an inability to obtain 'sorted' and evaluated information on which they can make modernization decisions. Small rural firms generally are unaware of industry best practice and of ways to improve incrementally. 'They don't know what they don't know' (Rosenfeld, 1992, p. 129).

Recent empirical research is demonstrating that, in some cases, a peripheral location is less of a hindrance to innovative activity than previously had been thought. The conventional wisdom has been that rural areas and places distant from metropolitan cores are disadvantaged with respect to innovation (Ewers and Wettmann, 1980). In support of this line of thinking, MacPherson's work in the USA has found that, in New York State, firms with greater access to external sources of expertise are more likely to utilize such sources of innovation (MacPherson, 1997). However, research in the Netherlands and in the UK, by contrast, has had difficulty identifying significant gaps between core and peripheral regions (Keeble, 1997). The reasons appear to be related to two interrelated things: first, *interfirm differences in the degree to which active, extroverted behavior takes place* and, second, the *'technical culture' of the locality or region*, much of which is summed up by the characteristics of technologically successful regions discussed above (Malecki, 1997; Sweeney, 1991). To some extent, these influences may also operate in more peripheral regions.

In peripheral places, firms must be outward-looking in order to succeed, seeking knowledge from outside the local region, and they must develop markets outside the restricted local area. The disadvantage of peripheries derives from the obvious obstacle, relative to, e.g., more central areas, is access to suppliers and customers, information, and labor (Vaessen and Wever, 1993; see also Crevoisier, Chapter 2, *this volume*). In other words, the network of successful firms in peripheral areas must be nonlocal to a considerable extent. These nonlocal networks frequently center on contacts made by owner-managers in previous employment (Malecki and Veldhoen, 1993; Vaessen and Wever, 1993). The difficulty of assembling the appropriate combination of hardware, soft factors, and conventions to support innovation is why few peripheral regions can be called adopters, and even fewer are adapters.

Challenges for Geographical Research on Technological Change

This chapter has proposed a framework according to which technological development takes place in spatial innovation systems. This framework is helpful in putting recent interdisciplinary research on technological development into a perspective that is meaningful to research in economic geography. As the discussion shows, economic geographers have gained a great deal of understanding on the local and regional systems of innovation. We understand that learning is critical, that networks help learning to occur, and that external connections are at least as important as those internal to a locality or region. What remain to be further explored in these systems (among other things) are the 'soft' sides of them, the 'not strictly economic' aspects of interaction (embeddedness, institutions, conventions and 'all that'), which we understand are often the crucial ones but are still in search of fitting concepts and their related contents.

Another terrain in these systems that remains relatively little understood are the connections between regional systems. It is widely acknowledged that local as well as nonlocal sources of innovative activity are crucial for innovations to occur but we are just beginning to understand the details related to the cofunctioning of proximity versus distance effects in various sorts of innovation (Gertler, 1995; Oinas and Virkkala, 1997). Several questions remain. We do not know very much about how successful firms build their local and extra-local networks of contacts. Are local relations first? Or do they follow the relations to other regions? And does it matter which relations are first, as long as the firm can survive until the 'appropriate' network is assembled? Does a region's success depend on a specific degree of 'globalness' in its firms' networks? Or are the local relations really the most important? We have some hints about these matters. For example, the necessary progression by a firm from a technological focus to markets (Roberts, 1990) typically coincides with a shift in linkages from local to national and international markets (Autio, 1994; Christensen, 1991; Christensen and Lindmark, 1993). Similarly, the personal networks of owner-managers vary according to the strategy chosen for the firm (Ostgaard and Birley, 1994).

We suspect that it is a very complex combination of the two sets of connections. And we suspect that only more research (including that of a qualitative nature) that probes the nature of the various types of significant relations between firms and their environments will uncover what is not known.

The chapters which follow add to our conceptual and empirical knowledge of technological change, and move forward our collective understanding of some of these problematic issues.

Note

1   They refer to 'strong institutional presence' but in order to avoid confusion with 'institutions' as 'rules', 'customs' or 'practices' as in much of the contemporary institutionalist literature, we use 'organisations' here.

References

Amendola, M. and Bruno, S. (1990), 'The Behaviour of the Innovative Firm: Relations to the Environment', *Research Policy*, vol. 19, pp. 419-433.

Amin, A. (1994), 'The Difficult Transition from Informal Economy to Marshallian Industrial District', *Area*, vol. 26, pp. 13-24.

Amin, A. and Thrift, N. (1992), 'Neo-Marshallian Nodes in Global Networks', *International Journal of Urban and Regional Research*, vol.16, pp. 571-587.

Amin, A. and Thrift, N. (1993), 'Globalization, Institutional Thickness and Local Prospects', *Revue d'Economie Régionale et Urbaine*, no. 3, pp. 405-427.

Amin, A. and Thrift, N. (1994a), 'Living in the Global', in A. Amin and N. Thrift (eds), *Globalization, Institutions and Regional Development in Europe*, Oxford University Press, Oxford, pp. 1-22.

Amin, A. and Thrift, N. (1994b), 'Holding Down the Global', in A. Amin and N. Thrift (eds), *Globalization, Institutions, and Regional Development in Europe*, Oxford University Press, Oxford, pp. 257-260.

Amin, A. and Thrift, N. (1995), 'Globalisation, Institutional "Thickness" and the Local Economy', in P. Healey, S. Cameron, S. Davoudi, S. Graham and A. Madani-Pour (eds), *Managing Cities: The New Urban Context*, John Wiley and Sons, London, pp. 91-108.

Arcangeli, F. (1993), 'Local and Global Features of the Learning Process', in M. Humbert (ed.), *The Impact of Globalisation on Europe's Firms and Industries*, Pinter, London, pp. 34-41.

Archibugi, D. and Michie, J. (1995), 'The Globalisation of Technology: A New Taxonomy', *Cambridge Journal of Economics*, vol. 19, pp. 121-140.

Archibugi, D. and Michie, J. (1997), 'Technological Globalisation or National Systems of Innovation', *Futures*, vol. 29, pp. 121-137.

Asheim, B.T. (1996), 'Industrial Districts as "Learning Regions": A Condition for Prosperity', *European Planning Studies*, vol. 4, pp. 379-400.

Autio, E. (1994), 'New, Technology-Based Firms as Agents of R&D and Innovation: An Empirical Study', *Technovation*, vol. 14, pp. 259-273.

Bahrami, H. and Evans, S. (1995), 'Flexible Re-cycling and High-Technology Entrepreneurship', *California Management Review*, vol. 37 (3), pp. 62-89.

Becattini, G. (1991), 'The Industrial District as a Creative Milieu', in G. Benko and M. Dunford (eds), *Industrial Change and Regional Development*, Belhaven, London, pp. 102-114.

Best, M.H. (1990), *The New Competition: Institutions of Industrial Restructuring*, Polity Press, Oxford.

Bull, A.C., Pitt, M. and Szarka, J. (1991), 'Small Firms and Industrial Districts: Structural Explanations of Small Firm Viability in Three Countries', *Entrepreneurship and Regional Development*, vol. 3, pp. 83-99.

Camagni, R. (1995), 'Global Network and Local Milieu: Towards a Theory of Economic Space', in S. Conti, E.J. Malecki, and P. Oinas (eds), *The Industrial Enterprise and Its Environment: Spatial Perspectives*, Avebury, Aldershot, pp. 195-214.

Cantwell, J. (1995), 'The Globalisation of Technology: What Remains of the Product Cycle Model?', *Cambridge Journal of Economics*, vol. 19, pp. 155-174 .

Carlsson, B. (1994), 'Technological Systems and Economic Performance', in M. Dodgson and R. Rothwell (eds), *The Handbook of Industrial Innovation*, Edward Elgar, Aldershot, pp. 13-24.

Castells, M. and Hall, P. (1994), *Technopoles of the World: The Making of 21st Century Industrial Complexes*, Routledge, London.

Christensen, P.R. (1991), 'The Small and Medium-Sized Exporters' Squeeze: Empirical Evidence and Model Reflections', *Entrepreneurship and Regional Development*, vol. 3, pp. 49-65.

Christensen, P.R. and Lindmark, L. (1993), 'Location and Internationalization of Small Firms', in L. Lundqvist and L.O. Persson (eds), *Visions and Strategies in European Integration*, Springer-Verlag, Berlin, pp. 131-151.

Conti, S. (1993), 'The Network Perspective in Geography: Towards a Model', *Geografiska Annaler*, vol. 75B, pp. 115-130.

Cooke, P. (1996), 'The New Wave of Regional Innovation Networks: Analysis, Characteristics and Strategy', *Small Business Economics*, vol. 8, pp. 159-171.

Cooke, P. (1997), 'Regions in a Global Market: The Experiences of Wales and Baden-Württemberg', *Review of International Political Economy*, vol. 4, pp. 349-381.

Cooke, P. and Morgan, K. (1994a), 'The Creative Milieu: A Regional Perspective on Innovation', in M. Dodgson and R. Rothwell (eds), *The Handbook of Industrial Innovation*, Edward Elgar, Aldershot, pp. 25-32.

Cooke, P. and Morgan, K. (1994b), 'The Regional Innovation System in Baden-Württemberg', *International Journal of Technology Management*, vol. 9, pp. 394-429.

Crewe, L. (1996), 'Material Culture: Embedded Firms, Organizational Networks and the Local Economic Development of a Fashion Quarter', *Regional Studies*, vol. 30, pp. 257-272.

Curran, P. and Blackburn, R. (1994), *Small Firms and Local Economic Networks: The Death of the Local Economy?*, Paul Chapman, London.

Dicken, P. and Thrift, N. (1992), 'The Organization of Production and the Production of Organization: Why Business Enterprises Matter in the Study of Geographical Industrialization', *Transactions of the Institute of British Geographers*, vol. NS 17, pp. 279-291.

Dicken, P., Forsgren, M. and Malmberg, A. (1994), 'The Local Embeddedness of Transnational Corporations', in A. Amin and N. Thrift (eds), *Globalization, Institutions, and Regional Development in Europe*, Oxford University Press, Oxford, pp. 23-45.

Dosi, G. (1988), 'Sources, Procedures, and Microeconomic Effects of Innovation', *Journal of Economic Literature*, vol. 26, pp. 1120-1171.

Dunning, J.H. (1994), 'Multinational Enterprises and the Globalization of Innovatory Capacity', *Research Policy*, vol. 23, pp. 67-88.

Eliasson, G. (1996), 'Spillovers, Integrated Production and the Theory of the Firm', *Journal of Evolutionary Economics*, vol. 6, pp. 125-140.

Estimé, M.-F., Drilhon, G. and Julien, P.-A. (1993), *Small and Medium-sized Enterprises: Technology and Competitiveness*, OECD, Paris.

Ewers, H.-J. and Wettmann, R.W. (1980), 'Innovation-oriented Regional Policy', *Regional Studies*, vol. 14, pp. 161-179.

Florida, R. (1995a) 'The Industrial Transformation of the Great Lakes Region', in P. Cooke (ed.), *The Rise of the Rustbelt*, UCL Press, London, pp. 163-176.

Florida, R. (1995b), 'Toward the Learning Region', *Futures*, vol. 27, pp. 527-536.

Freeman, C. (1987), *Technology Policy and Economic Performance*. Pinter Publishers, London.

Gertler, M.S. (1995), '"Being There": Proximity, Organization, and Culture in the Development and Adoption of Advanced Manufacturing Technologies', *Economic Geography*, vol. 71, pp. 1-26.

Grabher, G. (1993), 'The Weakness of Strong Ties: The Lock-in of Regional Development in the Ruhr Area', In G. Grabher (ed.), *The Embedded Firm: On the Socioeconomics of Industrial Networks*, Routledge, London, pp. 255-277.

Granovetter, M. (1985), 'Economic Action and Social Structure: The Problem of Embeddedness', *American Journal of Sociology*, vol. 91, pp. 481-510.

Grotz, R. and Braun, B. (1993), 'Networks, Milieux and Individual Firm Strategies: Empirical Evidence of an Innovative SME Environment', *Geografiska Annaler*, vol. 75B, pp. 149-162.

Hardill, I., Fletcher, D. and Montagné-Villette, S. (1995), 'Small firms' "Distinctive Capabilities" and the Socioeconomic Milieu: Findings from Case Studies in Le Choletais (France) and the East Midlands (UK)', *Entrepreneurship and Regional Development*, vol. 7, pp. 167-186.

Harrison, B. (1992), 'Industrial Districts: Old Wine in New Bottles?', *Regional Studies*, vol. 26, pp. 469-483.

Harrison, B. (1994), *Lean and Mean: The Changing Landscape of Corporate Power in the Age of Flexibility*, Basic Books, New York.

Harrison, B., Kelley, M.R. and Gant, J. (1996), 'Innovative Firm Behavior and Local Milieu: Exploring the Intersection of Agglomeration, Firm Effects, and Technological Change', *Economic Geography*, vol. 72, pp. 233-258.

Hassink, R. (1997), 'What Distinguishes "Good" from "Bad" Industrial Agglomerations?', *Erdkunde*, vol. 51, pp. 2-11.

Herrigel G. (1993), 'Large Firms, Small Firms, and the Governance of Flexible Specialization: The Case of Baden Württemberg and Socialized Risk', in B. Kogut (ed.), *Country Competitiveness: Technology and the Organizing of Work*, Oxford University Press, New York, pp. 15-35.

Hobday, M. (1994), 'Innovation in Semiconductor Technology: The Limits of the Silicon Valley Model', in M. Dodgson and R. Rothwell (eds), *The Handbook of Industrial Innovation*, Edward Elgar, Aldershot, pp. 154-168.

Johannessen, J.-A., Dolva, J.O. and Olsen, B. (1997), 'Organizing Innovation: Integrating Knowledge Systems', *European Planning Studies*, vol. 5, pp. 331-349.

Johannisson, B., Alexanderson, O., Nowicki, K. and Senneseth, K. (1994), 'Beyond Anarchy and Organization: Entrepreneurs in Contextual Networks', *Entrepreneurship and Regional Development*, vol. 6, pp. 329-356.

Julien, P.-A. (1995), 'Economic Theory, Entrepreneurship and New Economic Dynamics', in S. Conti, E.J. Malecki, and P. Oinas (eds), *The Industrial Enterprise and Its Environment: Spatial Perspectives*, Avebury, Aldershot, pp. 123-142.

Keeble D. (1997), 'Small Firms, Innovation and Regional Development in Britain in the 1990s', *Regional Studies*, vol. 31, pp. 281-293.

Kelley, M.R. and Brooks, H. (1992), 'Diffusion of NC and CNC Machine Tool Technologies in Large and Small Firms', in R.U. Ayres, W. Haywood and I. Tchijov (eds), *Computer Integrated Manufacturing*, vol. III. *Models, Case Studies, and Forecasts of Diffusion*, Chapman and Hall, London, pp. 117-135.

Kim, L. (1997), *Imitation to Innovation: The Dynamics of Korea's Technological Learning*, Harvard University Press, Boston.

Leonard-Barton, D. (1995), *Wellsprings of Knowledge*, Harvard Business School Press, Boston.

Lipietz, A. (1993), 'The Local and the Global: Regional Individuality or Interregionalism?', *Transactions, Institute of British Geographers*, vol. NS 18, pp. 8-18.

Luger, M.I. and Goldstein, H.A. (1991), *Technology in the Garden: Research Parks and Regional Economic Development*, University of North Carolina Press, Chapel Hill.

MacPherson, A. (1992), 'Innovation, External Technical Linkages and Small-Firm Commercial Performance: An Empirical Analysis from Western New York', *Entrepreneurship and Regional Development*, vol. 4, pp. 165-183.

MacPherson, A.D. (1995), 'Product Design Strategies amongst Small- and Medium-Sized Manufacturing Firms: Implications for Export Planning and Regional Economic Development', *Entrepreneurship and Regional Development*, vol. 7, pp. 329-348.

MacPherson, A. (1997), 'The Role of Producer Service Outsourcing in the Innovation Performance of New York State Manufacturing Firms', *Annals of the Association of American Geographers*, vol. 87, pp. 52-71.

Maillat, D. (1995), 'Territorial Dynamic, Innovative Milieus and Regional Policy', *Entrepreneurship and Regional Development*, vol. 7, pp. 157-165.

Maillat, D. (1996), 'Regional Productive Systems and Innovative Milieux', in OECD, *Networks of Enterprises and Local Development*, OECD, Paris, pp. 67-80.

Maillat, D. and Vasserot, J.-Y. (1988), 'Economic and Territorial Conditions for Indigenous Revival in Europe's Industrial Regions', in P. Aydalot and D. Keeble (eds), *High Technology Industry and Innovative Environments: The European Experience*, Routledge, London, pp. 163-183.

Malecki, E.J. (1997), *Technology and Economic Development: The Dynamics of Local, Regional and National Competitiveness*, 2nd edition, Addison Wesley Longman, London.

Malecki, E.J. and Tödtling, F. (1995), 'The New Flexible Economy: Shaping Regional and Local Institutions for Global Competition', in C.S. Bertuglia, M.M. Fischer and G. Preto (eds), *Technological Change, Economic Development and Space*, Springer-Verlag, Berlin, pp. 276-294.

Malecki, E.J. and Veldhoen, M.E. (1993), 'Networks Activities, Information and Competitiveness in Small Firms', *Geografiska Annaler*, vol. 75B, pp. 131-147.

Malmberg, A. and Maskell, P. (1997a), 'Towards an Explanation of Regional Specialisation and Industry Agglomeration', *European Planning Studies*, vol. 5, pp. 25-41.

Malmberg, A. and Maskell, P. (1997b), 'Towards an Explanation of Regional Specialisation and Industry Agglomeration', in H. Eskelinen (ed.), *Regional Specialisation and Local Environment – Learning and Competitiveness*, NordREFO, Copenhagen, pp. 14-39.

Maskell, P. and Malmberg, A. (1999), 'Localised Learning and Industrial Competitiveness', *Cambridge Journal of Economics, forthcoming.*

Massey, D., Quintas, P. and Wield, D. (1992), *High Tech Fantasies: Science Parks in Society, Science and Space*, Routledge, London.

Micklethwait, J. (1997), 'Future Perfect? A Survey of Silicon Valley', *The Economist*, 29 March.

Mody, A., Suri, R. and Tatikonda, M. (1995), 'Keeping Pace with Change: International Competition in Printed Circuit Board Assembly', *Industrial and Corporate Change*, vol. 4, pp. 583-613.

Mohan-Neill, S.I. (1995), 'The Influence of a Firm's Age and Size on Its Environmental Scanning Activities', *Journal of Small Business Management*, vol. 33 (4), pp. 10-21.

Morgan, K. (1996), 'Learning-by-Interacting: Inter-Firm Networks and Enterprise Support', in OECD, *Networks of Enterprises and Local Development*, OECD, Paris, pp. 53-66.

Mueller, F. and Loveridge, R. (1995) 'The "Second Industrial Divide"? The Role of the Large Firm in the Baden-Württemberg Model', *Industrial and Corporate Change*, vol. 4, pp. 555-582.

Nelson, R.R. (1996), 'The Evolution of Comparative or Competitive Advantage: a preliminary report on a study', *Industrial and Corporate Change*, vol. 5, pp. 597-617.

OECD (1996), *Networks of Enterprises and Local Development: Competing and Co-operating in Local Productive Systems*, Organisation for Economic Co-operation and Development, Paris.

Ohmae, K. (1995), *The End of the Nation State: The Rise of Regional Economies*, Free Press, New York.

Oinas, P. (1997), 'On the Socio-Spatial Embeddedness of Business Firms', *Erdkunde*, vol. 51 (1), pp. 23-32.

Oinas, P. and Van Gils, H. (1998), '"Learning Regions" and Firms – Where is the Learning?', *Mimeo*, Economic Geography/Applied Economics, Erasmus University Rotterdam, The Netherlands.

Oinas, P. and Virkkala, S. (1997), 'Learning, Competitiveness and Development. Reflections on the Contemporary Discourse on "Learning Regions"', in H. Eskelinen (ed.), *Regional Specialisation and Local Environment - Learning and Competitiveness*, NordREFO, Copenhagen, pp. 263-277.

Ostgaard, T.A. and Birley, S. (1994), 'Personal Networks and Firm Competitive Strategy – A Strategic or Coincidental Match?', *Journal of Business Venturing*, vol. 9, pp. 281-305.

Patel, P. (1995), 'Localised Production of Technology for Global Markets', *Cambridge Journal of Economics*, vol. 19, pp. 141-153.

Perrow, C. (1992), 'Small-Firm Networks', in N. Nohria and R.G. Eccles (eds), *Networks and Organizations*, Harvard Business School Press, Boston, pp. 445-470.

Porter, M.E. (1990), *The Competitive Advantage of Nations*, Free Press, New York.

Porter, M.E. and Wayland, R.E. (1995), 'Global Competition and the Localization of Competitive Advantage', in H. Thorelli (ed.), *Advances in Strategic Management*, vol. 11, part A: *Integral Strategy: Concepts and Dynamics*, JAI Press, Greenwich, CT, pp. 63-105.

Quintas, P., Wield, D. and Massey, D. (1992), 'Academic-Industry Links and Innovation: Questioning the Science Park Model', *Technovation*, vol. 12, pp. 161-175.

Roberts, E.B. (1990), 'Evolving Toward Product and Market-Orientation: The Early Years of Technology-Based Firms', *Journal of Product Innovation Management*, vol. 7, pp. 274-287.

Rosenfeld, S. (1992), *Competitive Manufacturing: New Strategies for Rural Development*, Center for Urban Policy Research Press, Piscataway, NJ.

Rosenfeld, S.A. (1997), 'Bringing Business Clusters into the Mainstream of Economic Development', *European Planning Studies*, vol. 5, pp. 3-23.

Rothwell, R. (1992), 'Successful Industrial Innovation: Critical Factors for the 1990s', *R&D Management*, vol. 22, pp. 221-239.

Saxenian, A. (1994), *Regional Advantage*, Harvard University Press, Cambridge, MA.

Schienstock, G. (1996), 'Towards a New Technology and Innovation Policy', in O. Kuusi (ed.), *Innovation Systems and Competitiveness*, Taloustieto Oy, Helsinki, pp. 86-90.

Simmie, J. (ed.) (1997), *Innovation, Networks and Learning Regions?*, Jessica Kingsley, London.

Singh, M.S. (1995), 'Formation of Local Skills Space and Skills Networking: The Experience of the Electronics and Electrical Sector in Penang,' in B. van der Knaap and R. Le Heron (eds), *Human Resources and Industrial Spaces*, John Wiley, Chichester, pp. 197-226.

Sternberg, R. (1996), 'Technology Policies and the Growth of Regions: Evidence from Four Countries', *Small Business Economics*, vol. 8, pp. 75-86.

Stöhr, W.B. (1990), 'Synthesis', in W.B. Stöhr (ed.), *Global Challenge and Local Response*, Mansell, London, pp. 1-19.

Storper, M. (1993), 'Regional "Worlds" of Production: Learning and Innovation in the Technology Districts of France, Italy and the USA', *Regional Studies*, vol. 27, pp. 433-455.

Storper, M. (1995), 'The Resurgence of Regional Economies, Ten Years Later: The Region as a Nexus of Untraded Interdependencies', *European Urban and Regional Studies*, vol. 2, pp. 191-221.

Storper, M. (1997), *The Regional World*, Guilford, New York.

Storper, M. and Salais, R. (1992), 'The Four "Worlds" of Contemporary Industry', *Cambridge Journal of Economics*, vol. 16, pp. 169-193.

Storper, M. and Salais, R. (1997), *Worlds of Production*, Harvard University Press, Cambridge, MA.

Sweeney, G.P. (1991), 'Technical Culture and the Local Dimension of Entrepreneurial Vitality', *Entrepreneurship and Regional Development*, vol. 3, pp. 363-378.

Sweeney, G.P. (1996), 'Learning Efficiency, Technological Change and Economic Progress', *International Journal of Technology Management*, vol. 11, pp. 5-27.

Tödtling, F. (1995), 'The Innovation Process and Local Environment', in S. Conti, E.J. Malecki and P. Oinas (eds), *The Industrial Enterprise and Its Environment: Spatial Perspectives*, Avebury, Aldershot, pp. 171-193.

Tödtling, F. and Sedlacek, S. (1997), 'Regional Economic Transformation and the Innovation System of Styria', *European Planning Studies*, vol. 5, pp. 43-63.

Tsipouri, L.J. (1991), 'The Transfer of Technology Issue Revisited: Some Evidence from Greece', *Entrepreneurship and Regional Development*, vol. 3, pp. 145-157.

Turok, I. (1993), 'Inward Investment and Local Linkages: How Deeply Embedded is Silicon Glen?', *Regional Studies*, vol. 27, pp. 401-417.

Tushman, M.L. and Rosenkopf, L. (1992), 'Organizational Determinants of Technological Change: Toward a Sociology of Technological Evolution', in B.M. Staw and L.L. Cummings (eds), *Research in Organizational Behavior*, vol. 14, JAI Press, Greenwich, CT, pp. 311-347.

Utterback, J.M. (1994), *Mastering the Dynamics of Innovation*, Harvard Business School Press, Boston.

Uzzi, B. (1996), 'The Sources and Consequences of Embeddedness for the Economic Performance of Organizations: The Network Effect', *American Sociological Review*, vol. 61, pp. 674-698.

Uzzi, B. (1997). 'Social Structure and Competition in Interfirm Networks: The Paradox of Embeddedness', *Administrative Science Quarterly*, vol. 42, pp. 35-67.

Vaessen, P. and Keeble, D. (1995), 'Growth-oriented SMEs in Unfavourable Regional Environments', *Regional Studies*, vol. 29, pp. 489-505.

Vaessen, P. and Wever, E. (1993), 'Spatial Responsiveness of Small Firms', *Tijdschrift voor Economische en Sociale Geografie*, vol. 84, pp. 119-131.

Zukin, S. and DiMaggio, P. (1990), 'Introduction', in S. Zukin and P. DiMaggio (eds), *Structures of Capital: The Social Organization of the Economy*. Cambridge University Press, Cambridge, pp. 1-36.

2 Globalisation and Industrial Competitiveness: The Process and Consequences of Ubiquitification

Peter Maskell

> Every locality has incidents of its own which affect in various ways the methods of arrangement of every class of business that is carried on in it ... The tendency to variation is the chief cause of progress (Alfred Marshall, 1890).

Introduction

The aim of this predominantly conceptual chapter is to investigate how the process of globalisation changes the way firms compete and interact. Combining traditional location theory with the *resource-based view* of the firm the chapter analyses how firms in high-cost countries and regions meet the new competitive challenge from cost-wise more favourably located competitors by increasing the rate of innovation at all levels. This in turn calls for inter-organisational co-operation as innovation is predominantly an interactive process (Rosenberg, 1972; Håkansson, 1987, 1989; Freeman, 1991). The ability of firms to select and connect to relevant local, domestic or foreign partners is thus becoming increasingly critical.

The cost of obtaining the information needed for choosing new suitable partners rises as the cultural distance increases (Hallén and Johanson, 1985), thus creating strong incentives for firms to engage with *local* partners whenever possible. It is also on the local level, where firms share the same values, background and understanding of technical and commercial problems, that the exchange of tacit knowledge is faced with the fewest barriers (von Hippel, 1994). The tacitness of localised learning-by-interacting furthermore prevents dissemination of the resulting knowledge to outsiders, thereby adding to the sustainability of the firm's competitive position. The chapter discusses how being embedded in a mesh of local connections helps firms survive and thrive.

The chapter is structured along the following lines. In the next section the spatially unequal consequences of the process of globalisation will be considered. In the following section the standard Weberian framework is supplemented with notions of competence and capabilities from modern theories of business economics. Section four contemplates how firms might escape the devastating effects of 'ubiquitification' by making connections in order to enhance their knowledge-creation. In section five the problems of competitiveness stemming from codification of former tacit knowledge are considered. The final section muses on how competitiveness is achieved when firms embed in a local environment by utilising mainly local connections. In conclusion, it is argued that the formation of global markets has not erased the spatial distinctions that make firms differ in strategy and performance.

Globalisation and the Location of Economic Activity

Recently a number of studies have pointed out how globalisation is not as modern a phenomenon as the contemporary public debate might make us believe. Only during the late 1990s have export rates and foreign direct investments surpassed the level attained eighty years earlier (Feis, 1934; Bairoch, 1996; Maddison, 1991) – i.e., before the First World War signalled an abrupt decrease in what Cournot once called 'the territory of which the parts are so united by the relations of unrestricted commerce that prices there take the same level throughout, with ease and rapidity' (Cournot (1838)1927, pp. 51-52). The recovery of most product markets was very slow and another world war long gone before the territory of unrestricted commerce again advanced beyond the national borders. The current development is thus picking up on an old trend and is driven by the same desire to obtain 'the general opulence' described by Adam Smith ((1776) 1979, p. 117).[1]

The rediscovery of the magnitude of globalisation at the beginning of the century reminds us that a transnational market is a delicate creature, its existence always set to be discontinued by some future event. Not only military action but also less dramatic incidents such as congestion, infrastructure under-investments, and trade wars may raise the relative cost of transporting a tangible or intangible good and thereby reverse the process of globalisation. Even if no such event occurs, any reduction in the cost of producing a good will of course make its movement over space *relatively* more expensive. Continued globalisation thus hinges on the prospect of matching any future cost-saving improvement in production with an equally cost-saving improvement in transportation and communication. It is still an open question how much longer such matching improvements can be found especially regarding tangibles.

Globalisation has, however, undoubtedly become 'the favorite business buzzword of the '90s' (Fleenor, 1993, p. 7) and also the subject of escalating academic interest: The number of books with 'global' in the title rose from 82 in the 1950s, through 303 and 1,766 in the 1960s and 1970s, reaching 4,496 in the 1980s (Worthington, 1993). The equivalent number for the 1990s will count five digits. But in spite of this immense interest we still lack a commonly accepted definition of globalisation. Some argue that the phrase should be reserved for describing the final stage in the reconfiguration of economic interaction, beginning with 'transnational' activities over 'multinational' and 'international' operations eventually reaching the crowning stage of 'globalisation', where the predominant parts of economic life are generated and utilised between actors in different countries (Vickery, 1996; Holstein, 1990). Others have chosen to use the concept of 'globalisation' to describe the process rather than the outcome (Hu, 1992; Dicken, 1994; Gordon, 1995; Wade, 1996). The author of this chapter belongs to the latter congregation.

Regardless of the widespread reservations and mixed sentiments stemming from the ambiguous ways in which the term globalisation is applied, no one can doubt that international economic interaction has increased in recent decades. Investments and innovations in transport and communication systems, and the success of governmental efforts to remove most former barriers to trade (GATT, WTO), have jointly made national frontiers more porous. One frequently used indicator to illustrate the phenomenon is the increased rates of exports and imports in relation to the total production of most countries (Table 2.1).

The present process of globalisation shares important features with the pre-WW1 'empire' model of globalisation, but it also contains genuinely novel elements as the new 'ethos of globalisation' makes many former national loyalties dwindle. Consumers increasingly choose commodities primarily on the basis of price, quality and reputation (brand), making considerations of producer nationality of secondary importance.

Table 2.1   Foreign trade as a percentage of production in OECD countries, 1970-1992

	1970	1981	1992
Australia	29	39	45
Austria	48	60	80
Belgium	85	128	147
Canada	50	61	85
Denmark	80	92	103
Finland	56	57	70
France	30	45	60
Germany	31	46	56
Italy	29	43	47
Japan	12	17	17
Netherlands	82	112	133
New Zealand	64	72	77
Norway	77	74	84
Spain	15	22	51
Sweden	53	68	81
United Kingdom	30	46	62
United States	11	18	29
OECD Total	23	33	42

Note: trade = exports + imports
Source: OECD STAN Database, 1995.

International business standards, insurance systems and banking facilities for swift and secure transfer of payments have also contributed to a considerable transformation and growth in transnational interaction on most business-to-business markets.

The surge in the number, scale and scope of cross-border inter-firm collaborations reinforces the effect obtained by the expansion of internationally operating companies (Dicken, 1992). Furthermore, the globalisation of capital markets now influences the financial policies of even the world's strongest economies, but information problems and monitoring costs nevertheless still steer most savings into domestic investments.

The latter point may make one assume that the founder of international trade theory, David Ricardo (1772-1823), was basically right when he once stated:

> Experience, however, shews that the fancied or real insecurity of capital, when not under the immediate control of its owner, together with the natural disinclination which every man has to quit the country of his birth and connexions, and intrust himself with all his habits fixed, to a strange government and new laws, check the emigration of capital. These feelings, which I should be sorry to see weakened, induce more men of property to be satisfied with a low rate of profits in their own country, rather than seek a more advantageous employment for their wealth in foreign nations (Ricardo, 1817, pp. 161-162).

But in contrast to the situation when Ricardo addressed these problems 180 years ago and until very recently, the new international institutional arrangements and the novel ethos of globalisation make few contemporary capitalists feel satisfied with lower domestic profits if the same commercial risk yields higher returns at some foreign destination (Vickery, 1996).

At a more fundamental level the genuine novel feature of the present process of globalisation is the promises it contains for the economically less developed parts of the globe (Ernst, 1981). Many basic features of inequality are certainly still at work (Bairoch and Kozul-Wright, 1996; Bienefeld, 1994), but any crude centre-periphery model of global economic development is, nevertheless, getting less and less valid as some of the former third world countries push ahead, and as it is becoming increasingly obvious that endogenous factors (including corruption, unproductive investments, conspicuous consumption, and ethnic tension) are at least as important for curtailing economic growth as any post-imperialistic dominance exercised from actors in the first world.

The growth in trade between OECD countries and countries outside the OECD area reveal that an important transmutation has taken place in the global economy. For perhaps the first time ever firms in less developed counties can access the better part of the world's markets on almost equal terms and acquire some of the most advanced technologies available while simultaneously benefiting from knowledge of others' successful organisational designs and marketing strategies.

These genuine new opportunities for domestic or foreign firms in some (but clearly not all) of the low-cost countries of the world, has sometimes occasioned amazingly high rates of growth in these areas. In the longer run modifying factors will, of course, come into play: the surplus on the balance of trade alone will eventually lead to an appreciation of the national currencies of the current low-cost nations which, ceteris paribus, will bring about a more homogenous cost regime. The current discrepancies are, however, of a magnitude that make any imminent levelling of labour cost across countries highly unlikely. Global cost differences are here to stay for a considerable time.

Little wonder that fears have been voiced regarding the flip side of the coin: the possible damaging consequences of the globalisation process for the high-cost regions of Europe, Japan and North America (Reich, 1990; Rodrik, 1997).[2] When globalisation goes too far, it is argued, firms will either emigrate or evaporate, leaving the old industrial countries in a slowly more desolate and jobless state. The widespread anxiety regarding the anticipated sweeping relocation of economic activity might justify probing a little further into industrial location theory.

Location Theory and the Resource-based View of the Firm

Traditionally a distinction has been made (Weber, 1909)[3] between on the one hand the factors of economic importance for the operation of a firm, for which the costs differ significantly between locations, and on the other hand the so-called ubiquitous materials: the ones used in production of a commodity, and which in practice are available everywhere at more or less the same cost. Weber stated that:

> As regards the nature of the material deposits, some materials employed in industry appear everywhere; they are, for practical purposes, put at our disposal by nature without regard for location. When the whole earth is considered this actually holds true only in the case of air; but when more limited regions are considered, this holds true for many other things ... Such materials will be called 'ubiquities' (Weber, 1909, p. 51, here in the translation by Friedrich, 1929, pp. 50-51).[4]

Weber used the distinction to determine the degree of market-pull on the location of industries: the larger an element of ubiquities in the final product, the stronger would the potential savings in transportation cost pull the industry producing such products away from the supplies of raw material and towards a location near their customers.

The Weberian distinction still holds, even though changes have occurred from time to time in the list of critically important *locational* factors. But for each and every locational factor for which former significance is shrinking, the position of some other factor must be rising. So when the locational factors of yesterday disappear from the list, a new list of the currently most prominent locational factors automatically takes shape.

Two processes traditionally determine the shifts in the relative importance of locational factors (Cunningham, 1902). Either the *demand* for a former important factor is weakened, for instance through some innovation in the production process, using other inputs than before, or changing the balance in which old inputs are being used. Alternatively the *supply* of localised input has changed: natural deposits are exhausted; new sources are discovered; labour becomes scarce; suppliers relocate; or the geographical patterns of demand have shifted.

As a repercussion of the process of globalisation a new, *third process* has now emerged, *actively converting* former localised inputs into becoming ubiquities: A large domestic market is no advantage when transport cost are negligible; when customers' local loyalties are dwindling, and when most custom barriers are eroded. Domestic suppliers of the most efficient production machinery are no longer an unquestioned blessing, when identical equipment is available world-wide, and at essentially the same cost. The omnipresence of organisational designs of proven value makes, furthermore, a long industrial track record less valuable. The process whereby globalisation create ubiquities can be termed 'ubiquitification' or – perhaps slightly less horrendously – 'the U-process'.

Hence, the relevance of the Weberian distinction has not tapered off as globalisation progressed. On the contrary, the Weberian distinction composes the pivotal linkage between locational theory and modern resource-based view of what constitutes a firm and make it succeed.[5] When the U-process reduces the heterogeneity of regions and nations competitiveness will be curtailed for all firms who did benefit from a previously favourable location, thus enhancing the competitiveness of firms located elsewhere.

Essentially the resource-based view of the firm tells us that competitiveness can only be built on heterogeneity: on firms having control over something wanted by others or by firms being able to do something that the competitors cannot do as well, as fast or as cheap.[6] Little progress would be made in a world of clones.

The resource-based view of the firm focuses on the stream of profitable services which derive from the human and other resources firms accrue (employ, buy, rent, inherit, etc.) and from the specific competencies they build when organising these resources (Barney, 1991). Many of a firm's assets, as shown on its balance sheet, do not add to the firms competitiveness because all its competitors have similar things. And what everyone has can never constitute an advantage. The telephone or air-conditioning system, the book-keeping department, the transport and maintenance functions, the parking space, the pencil and paper or the desks often belong to this category of 'pedestrian' resources that might be needed for day-to-day running of the firm, but which will usually *not* help it prosper (Montgomery, 1995).

The non-trivial resources consist of those that are both valuable and rare: that is, resources in high demand and with insufficient supply. Many firms have acquired long-lasting and strong competitive positions on their appropriation of basically unique resources, whether tangible (like a mineral deposit or a corner location) or intangible (like a patent, a fishing right or the control of the access to some customers). More firms, however, depend on their ability to *build* specific competencies based on the synergies between resources readily available to everyone on the market.[7]

Valuable built competencies will be imitated by competitors if not protected. But explicit safeguards such as legal protection through patent rights and other means often only offer the knowledge-owner a partial and temporary protection. Knowing that something is possible may provide enough guidelines to enable imitation in ways that do not necessarily violate any legal protection of original knowledge-owner. Every manmade object testifies by its mere existence that there must be some way to produce it. If the object is valuable, someone different from the initial producer will eventually come up with a solution, which might differ slightly from the original production process.

Outside certain industries (e.g., pharmaceuticals, electronics) legal protection is anyhow less prevalent. In these mainly non-patenting industries the durable competencies are often the outcome of a strenuous and time-consuming process of trial and error within the firm that step-by-step has made it better in performing some tasks in demand, or made it develop certain skills with a broad range of profitable applications. As such built competencies are a function of history it is usually 'impossible to simply copy best practice even when it is observed' (Rumelt *et al.*, 1991, p. 16). Competitors who try to imitate the proven successes of others are faced with the difficulty of identifying the precise procedures, organisational design and individual skills that constitute the revealed competence. Other imitative efforts might turn out to be commercially unattractive simply because of the time needed, as wonderfully shown in the classical dialogue between a British lord and his American visitor:

'How come you got such a gorgeous lawn?' 'Well, the quality of the soil is, I dare say, of the utmost importance.' 'No problem.' 'Furthermore, one does need the finest quality seed and fertilisers.' 'Big deal.' 'Of course, daily watering and weekly moving are jolly important.' 'No sweat, just leave it to me!' 'That's it.' 'No kidding?!' 'Oh, absolutely. There is nothing to it, old boy; just keep it up for five centuries.' (Dierickx and Cool, 1989, p. 1507).

Globalisation Obliterates Old Localised Capabilities

The many contributions to the resource-based view of the firm have mainly been preoccupied with resource-synergies, built competencies and other *internal* processes whereby firms create heterogeneity in order to obtain competitive advantages.

However, firms might, as all geographers know, just as well be *furnished* with sufficient heterogeneity by utilising specific factors in their *surroundings* not equally available to competitors located elsewhere (Foss, 1996). Firms without any major competencies or valuable resources can therefore survive and thrive if favourably located. A favourable location can, furthermore, enhance the competitiveness of firms already well-equipped with competencies of their own making.

In earlier times, the heterogeneity among localities was often caused by variations in the natural resource endowment. Today it is seldom the inherited natural properties but rather the *created localised capabilities* that establish the platform of heterogeneity on which the competitiveness of firms can be built or augmented. These localised capabilities include for instance the specific, but basically random, first-mover cost advantages; unlike patterns of demand and specialisation; disparate results of past investments; distinctive formal or informal institutional endowment, and dissimilar technological assets, all of which might make territorial entities differ from one another (Arthur, 1994).

As the process of globalisation involves the U-process – making formerly rare and valuable localised capabilities available to an increasing number of firms – the localised capability loses its importance. Firms, whose competitiveness depends on that capability, will be penalised on the market just as – on an aggregate level – the established patterns of regional or national specialisation will be jeopardised.

When Ubiquities are Created, Capabilities are Destroyed

As the process of globalisation gradually converts most natural properties and many created localised capabilities into ubiquities, the competitiveness of many firms will increasingly be associated with labour cost, as labour is still a largely immobile factor of production (Rasmussen, 1996). Even at a regional level labour is highly stationary in most developed countries – with the US as a remarkable exception. At the national level cross-country immobility characterises all kinds of labour, from the unskilled to the top positions on the labour market: American-based firms have almost exclusively American-born managers just as foreign directors in European or Japanese firms are 'as rare as British sumo wrestlers' (Wade, 1996, p. 79). Huge national differences in labour cost can thus persist while the prices of many other industrial inputs are becoming gradually more homogenous over space as globalisation progresses.

When the process of globalisation confront firms in high-cost environments with new competitors, located more favourably cost-wise, they react in different ways. Some raise their capital/labour-ratio through massive investments, thereby minimising the scope for low-wage competitors, but also minimising the firm's own ability to respond to changes in demand. Others out-source or relocate a part or most of their activities to low-cost areas, retaining only the least labour-intensive activities.

An increasing number of firms have to meet the challenges in a less habitual way by no longer chiefly aspiring to obtain competitiveness through cost reduction, but by generating entrepreneurial rents through enhanced knowledge creation (Nonaka, 1991; Spender, 1994; Gibbons *et al.*, 1994). The result of the process is reflected in the way each country and region is developing particular capabilities. As one of the growing number of empirical studies concludes:

> ... each country has developed a distinct model of specialization, concentrating its efforts in particular fields where world class capacities have often been developed ... There seems to be a specific advantage in a higher degree of specialization in technological fields, associated with the economics of scale and scope made possible at the national level. This advantage emerges regardless of the particular sector in which individual countries concentrate their efforts; in other words, for advanced countries being specialised appears to be even more important than choosing the 'right' field (Archibugi and Pianta, 1992, pp. 148 and 150).

Globalisation thus implies that the gist of industrial competitiveness in high-cost areas shifts step-by-step from static price competition towards dynamic improvement, benefiting firms that can create new knowledge more swiftly than their competitors.

This process of knowledge creation is never likely to run out of steam.[8] In spite of the accumulated efforts of humankind since the beginning of time, we probably still only know a tiny fragment of what we may find useful to know. And that again constitutes only a fraction of all there is to know (Lundvall, 1994). Common sense may tell us that the fraction of what is known will gradually increase over time. However, quite the contrary seems to be the case, not because of some global obliviousness, but because the unknown expands as we learn (Griliches, 1994). Knowledge creation is thus a Sisyphean process with no end or upper limit and is excellently suited as a basis for creating new sources of heterogeneity and subsequently for obtaining fresh competitive advantages.

Codification of Tacit Knowledge

Initially most pieces of knowledge probably appear in a form that is exclusively *tacit* (Polanyi, 1958, 1966): A person gets an idea or becomes aware of some hidden relationship or new opportunity (Cowan and Foray, 1997). Such purely tacit knowledge is at first accessible to the individual only, and a lot of new knowledge will remain that way. Sometimes, however, it is shared with others, who have the facilities to understand the idea and grasp its significance and implications. Still the knowledge remains in a mostly tacit form existing solemnly within this smaller group of persons, often sharing some common trait, which made the transmittal possible.

Over time, many such shared pieces of knowledge normally become gradually more *codified*. Codified knowledge can be communicated by symbols and language, and thus has the necessary features to be tradable (Dosi, 1988), if and when the sufficient market conditions occur. What is actually codified depends on the scope of the codification process, whether deliberate or not, and on the idiosyncrasies of the agents involved in the process.

Much former tacit knowledge has become codified over time through the increased understanding of its nature when used in different circumstances and by different persons or organisations. Sometimes, for instance, the new approach turns out under closer examination to be representing a general phenomenon, which gradually over the years might become formulated as a universal law or principle. More frequently, the new approach to a problem gets better understood by its use and refinement in practice. Gradually its constituting parts become identified as the new method is broken down to still more elementary segments. Codification is thus a mainly *unpremeditated consequence* of knowledge being used.

With each stage of unravelling and simplification, the description of the ingredients in the new approach becomes easier, and the prospects improve for communicating them to individuals unacquainted with the specifications of the original problem. Improving the prospect for exchanging codified knowledge does, however, not infer that the receiver of the knowledge can always use it immediately and effortlessly. Chess, for instance, is highly codified, but it nevertheless takes much effort to be able to play the game, even after having learned all the rules (Hatchuel and Weil, 1995). Codification thus makes exchange easier only in the respect that very little will have to be invested in the relationship between the present owner and the receiver of the knowledge, in order to convey its contents.

Though the reasons are not yet completely clear, it is also a common experience that some tacit knowledge is almost always required in order to use new codified knowledge. Nobody has, for instance, been able to operate a personal computer for the first time only be reading the manual, however carefully written. Only after a little practice with the key-board do the phrases and examples in the manual suddenly start making sense and the transmission of codified knowledge (the manual) to the new recipient can take place. The demand for a tacit knowledge-base prior to any successful transfer of codified knowledge might even help explain the depressing results of much development aid to countries and regions in the third world, while other parts of the world experience few difficulties in catching up with the technology used by the most advanced, once former barriers were eliminated.

Without codification the transfer of knowledge will typically take place by demonstration: the owner of the knowledge showing the novice how to behave. Successful codification thus infers a lasting reduction in the otherwise recurrent cost of communicating knowledge from one individual, department or organisation to another (Zander and Kogut, 1995). The more a firm is able to codify its tasks, the fewer resources are needed for instruction, guidance, training and supervision of the employees. When Henry Ford, for example, achieved a manifold increase in the productivity of some car-making processes in the beginning of the century, the main reason was not the invention of the conveyer-belt or any other technical innovation. His innovation was basically organisational – by conducting a meticulous codification of each task which was subsequently apportioned to an unskilled worker, needing only brief training to meet the limited demands of the task at hand (Ford, 1922; Hounshell, 1984).

This virtue of codified knowledge will in itself act as an incentive for still further codification. Technological progress is therefore to a large extent the result of an inter-linked process of knowledge creation and subsequent codification. Codification is thus at the heart of the whole philosophy of industrialisation.

Furthermore, and more importantly for the present discussion, codification of a piece of knowledge will enable its diffusion and subsequently undermine the ability of the owner to use it for the building of competitiveness. Once codified, knowledge is difficult to conceal and protect against imitation:

> Whether he communicates his results to a learned society, and leaves other to earn money by them, or applies them in practice himself (with or without the protection by a patent), they become in effect the property of the world almost at once. Even if he uses them in a 'secret process', enough information about them often leaks out to set others soon on the track near his own (Marshall (1919) 1927, p. 204).

So when former tacit knowledge is converted into a codified form, it rapidly becomes a ubiquity, accessible on the global market. For the possessor of former tacit knowledge, the immediate effect of codification is the same as all with all other former competencies that the firm might have benefitted from, but which has later become available to everyone: By becoming ubiquitous the knowledge loses its potential to contribute to the competitiveness of the firm.

The consequences of this phenomenon for the firms in high-cost areas of the world are dramatically increased as many other heterogeneous resources – the localised capabilities – are simultaneously turned into ubiquities by the process of globalisation. No firm exposed to international competition and located in a high-cost area can therefore depend solely on already fully codified knowledge. The more easily codified knowledge is accessed by everyone, the more crucial will tacit knowledge therefore become in sustaining or enhancing the competitive position of the firm.

Knowledge Creation and the Role of Geography

When intentionally creating new knowledge firms cultivate their ability to connect to others at the local, national and even global level (Ernst and Guerriei, 1997). These geographically expanding business networks enable firms to gain access to new sources of information, skills and production, thereby complementing the firm's own competencies and increasing the value of its assets.

For instance has DeBresson (1996) has shown that of the 1,641 major Canadian innovations from 1945-1970, less than 10 percent were the result of a firm's 'in-house' activities only. The rest involved as many as seven, and an average of four, different independent organisations. Historical studies (Rosenberg, 1972) have emphasised the importance of interaction between users and producers of machinery for the development of the textile industry in the US. In Sweden the Uppsala School, taking their starting point in the economics of industrial marketing, has in a number of studies empirically shown how competencies were enhanced through networks involving informal co-operation between firms (Håkansson, 1987, 1989). Others have demonstrated the importance of interaction between departments and with customers for innovation success and competitiveness (Freeman, 1982, 1991; Lundvall, 1985, 1988; Hagedoorn and Schakenraad, 1992), and the recent literature on innovation systems (Lundvall, 1992a, 1992b; Nelson, 1993; Freeman, 1995; Edquist, 1997) has this inter-activeness as its most basic building block.

Such an innovative process – requiring a high level of interaction, dialogue and exchange of information – may be conducted long-distance, but is often less expensive, more reliable and easier to conduct locally (David *et al.*, 1996). Contemporary empirical studies strongly support this view (Jaffe *et al.*, 1993; Malmberg, 1996). The clustering of inputs such as industrial and university R&D, agglomerations of manufacturing firms in related industries, and networks of business-service providers often create scale economies in the creation of knowledge, and facilitate the transfer of knowledge to the firms in the area (Patel and Pavitt, 1991; Patel, 1995).

Not only the practicality of being close to the relevant organisations is of importance. Also the ability to exchange otherwise purely internal information constitutes an important part of the competitive advantage of industrial agglomerations (Malmberg and Maskell, 1997). This ongoing process of geographically concentrated tacit knowledge-sharing and cross-fertilisation of ideas enhances knowledge creation at the level of the firm.

At the aggregate level of the locality the refinement of inter-organisational routines for knowledge transfer increases the efficiency of the economy by
lowering the total transaction cost (Sako, 1992). The capabilities of some areas
are evidently more predisposed than others to support special types of knowledge creation by establishing a specific 'culture' with routines and conventions
that make the economic system function without much fuss and with accordingly small transaction cost: the cost of persuading, negotiating, co-ordinating,
understanding and controlling each step in a transaction between firms
(Gambetta, 1988; Gertler, 1995; Maskell, 1998; Lorenz, 1998).

The socially-constructed framework, which enables firms to interchange
otherwise purely internal information, constitutes an important part of the total
set of capabilities which distinguish some nations or regions, and enhance the
competitiveness of the firms located there (Casson, 1993). When long-term
national or regional collective learning has taken place in a line of business, the
cost of using the market – as opposed to relying only on intra-firm activities
– diminishes to a point, where a territorial industrial configuration of only
small firms might become even more efficient than a configuration of larger
firms, burdened with the cost of internal control and measures against shirking
(Maskell, 1998).

Thus, a business environment that enhances trust will always make an
economic difference, but when the traditional, static, cost-related international
competition is superseded by competition based on dynamic improvements and
learning, the importance of such an environment has intensified dramatically.
Restrictions and barriers for conducting and engaging in an economically
beneficial learning-by-interaction are obviously diminished.

Learning-by-interaction is usually low-key, unprogrammed, incremental
refinement in the everyday operations of some aspect of the activities carried
out by the firms involved, which seem to be of paramount importance in
explaining the position of small high-cost nations in the international division
of labour (Maskell *et al.*, 1998). These nations typically specialise in low-tech
industries, where the competitiveness of the firms is maintained and augmented
by a constant flow of consecutive improvements, speedily disseminated to local
rivals, who will often further cultivate and improve the achievement. This
ongoing interaction represents a subtle but decisive enhancement of the
knowledge creation in all the firms of the area, simply because their co-location
enables them constantly to monitor each other better than any outsider. A local
club between rivals is constructed, beneficial to all its national, regional or
sometimes only local members. This localised learning is – almost by definition

– tacit in nature and so is much of the resulting improvements in the involved firms.

For firms impaired by the process of globalisation the knowledge-creation among proximate business partners becomes of crucial significance in rebuilding competitiveness. So when globalisation creates ubiquities and destroys capabilities, localised learning offers a way by which firms in high-cost environments might survive and thrive.

Conclusion

The focus in this chapter has been placed almost exclusively on one group of regions and nations: the ones with above average labour cost. The basic assumption is that the ongoing creation of a world market has required fundamental changes in the way in which high-cost nations and regions can defend and enhance their competitiveness. In these parts of the world long-term competitiveness is increasingly related to the ability of firms to upgrade their knowledge base and augment their performance continuously, rather than simply trying to obtain static efficiency through the identification and exploitation of cheap resources and economics of scale.

It has been argued that firms in high-cost areas of the world must be able to create or recreate valuable competencies at least as fast as they are destroyed by the U-process. Firms do this by making connections at all levels from the local to the global. However, inter-organisational co-operation is frequently cheaper and faster when it takes place at the local level than at great distance and when their tacitness make the results less prone to be imitated.

Some of the knowledge created through local inter-firm co-operation is furthermore embedded in the area's business culture, thus making it even more difficult for outsiders to imitate. Localised capabilities thereby translate into sustainable competencies, enabling firms to survive and thrive in spite of an unfavourable local cost-structure.

Neither regional nor national features and distinctions are completely washed away by the formation of global markets. Marshall's old statement, quoted at the very beginning of this chapter, still stands. Regardless of the U-process and many convergent developments in the world economy, economic progress is constituted by firms steadily embedded in specific territorial settings which contribute to the heterogeneity necessary for maintaining competitiveness.

Acknowledgement

The writing of the paper on which this chapter is based coincided in time with the author's participation in the completion of a book manuscript (Maskell *et al.*,1998) and some lines of the argument and some phrases from the book have made their way into the present chapter. Thanks to my co-authors for allowing this.

Notes

1 Economists working with the so-called increasing returns endogenous growth models (Romer, 1990; Sala-i-Martin, 1990; Rivera-Batiz and Romer, 1991; Grossman and Helpman, 1991, 1993) have identified a number of links between globalisation, trade liberalisation and economic growth (Baldwin and Forslid, 1996). See also Bretscher (1997).

2 It must be kept in mind that just as the group of 'high-cost areas' consist of a broad variety of spatial entities, so do the group of 'low-cost areas'. An informative and brief account of some of these differences is given by Kennedy (1993), while Zysman *et al.* (1996) focus on the countries of South-East Asia. Young (1994) shows that much of Asia's recent growth can be explained by factor accumulation and reallocation of resources.

3 Building on an existing tradition (Loria 1888, 1908; Maunier 1908), Alfred Weber later extended his original contribution (1909) to a 'locational theory under general and under capitalistic conditions' (1923). A short sketch of Weber's theory is given by Predöhl (1928). See also Jewkes (1933), Friis and Maskell (1980), Gregory (1982) and Malecki (1991).

4 In Weber's original wording, the definition goes like this: 'Die Materialien, die in der industrie verwandt werden, können für die großen Verhältnisse der Praxis angesehen, "überall" vorkommen, von der Natur also – eben im Großen angesehen – ohne Rücksicht auf den Ort zur Verfügung gestellt werden. Was ganz allgemein d.h. für alle Plätze der Erde ja eigentlich nur von der luft gilt, für enger begrenzte gegenden, die man zum Gebiet abgeschlossener Standortsbetrachtung machen kann, aber auch für viele andere Dinge … Solche Materialien mögen 'Ubiquitäten' heißen (Weber, 1909, p. 51).

5 The resource-based view of the firm is rooted in the seminal contribution of Penrose (1959). It was revived in the mid-1980s by Wernerfelt (1984), Rumelt (1984) and others, but it was Prahalad and Hamel's (1990) outstandingly successful article which more than anything sparked the interest of the business community and signalled a still-swelling stream of scientific contributions from

a gradually broadening group of scientific disciplines. Foss (1997) gives an overview and interpretation of the complex roots to the present resource-based view of the firm.

6   For those that worry about definitions, competitiveness means ' ... the ability of companies, industries, regions, nations or supernational areas to generate, while being and remaining exposed to international competition, relatively high factor income and factor employment levels on a sustainable basis' (Hatzichronoglou, 1996). The definition is inspired by Scott and Lodge (1985), Cohen and Zysman (1987), and Hatsopoulos *et al.* (1988) and it is now gaining ground also in international economic organisations like the OECD.

7   It is worth pointing out that, while resources can usually be bought at a price, there is no market for many built competencies like reputation, trustworthiness, reliability, or innovativeness, which nevertheless can augment the competitiveness of a firm by their continuous reuse when selling an expanding range of commodities.

8   The rate of innovation is, however, subject to the constraint of demand. No firm can survive if new innovations constantly make its old products obsolete before the initial investment in their development and production has been profitably recovered (Richardson, 1997).

References

Archibugi, D. and Pianta, M. (1992), *The Technological Specialization of Advanced Countries*, Kluwer, Dordrecht.

Arndt, C. and Hertel, T.W. (1997), 'Revisiting "The Fallacy of Free Trade"', *Review of International Economics*, vol. 5, pp. 221-229.

Arthur, W. B. (1994), *Increasing Returns and Path Dependence in the Economy*, University of Michigan Press, Ann Arbor.

Bairoch, P. (1996), 'Globalization Myths and Realities: One Century of External Trade and Foreign Investment', in R. Boyer and D. Drache (eds), *States against Markets: The Limits of Globalization*, Routledge, London, pp. 173-192.

Bairoch, P. and Kozul-Wright, R. (1996), *Globalization Myths: Some Historical Reflections on Integration, Industrialization and Growth in the World Economy*, United Nations Conference on Trade and Development (UNCTAD) discussion paper 113, UNCTAD, Geneva.

Baldwin, R. E. and Forslid, R. (1996), *Trade Liberalization and Endogenous Growth: A q-Theory Approach*, Discussion Paper 1397, Centre for Economic Policy Research (CEPR), London.

Barney, J. (1991), 'Firm Resources and Sustained Competitive Advantage', *The Journal of Management*, vol. 17, pp. 99-120.

Bienefeld, M. (1994), 'Capitalism and the Nation State in the Dog Days of the Twentieth Century', *Socialist Register*, vol. 30, pp. 94-129.

Bretscher, L. (1997), 'International Trade, Knowledge Diffusion and Growth', *The International Trade Journal*, vol. 11, pp. 327-348.

Casson, M. (1993), 'Cultural Determinants of Economic Performance', *Journal of Comparative Economics*, vol. 17, pp. 418-442.

Cohen, S.S. and Zysman, J. (1987), *Manufacturing Matters: The Myth of the Post-industrial Economy*, Basic Books, New York.

Cournot, A.A. (1838) *Recherches sur les principes mathématiques de la théorie des richesses* (Translated by Nathaniel T. Bacon (with a bibliography of mathematical economics by Irving Fisher); published 1897 (reprinted 1927) as *Researches into the Mathematical Principles of the Theory of Wealth by Augustin Cournot 1838*, Macmillan, New York).

Cowan, R. and Foray, D. (1997), 'The Economics of Codification and the Diffusion of Knowledge', *Industrial and Corporate Change*, vol. 6, pp. 595-622.

Cunningham, W. (1902), 'The Location of Industry', *The Economic Journal*, vol. XII, pp. 501-506.

David, P. A., Foray, D. and Dalle, J.-M. (1996), *Marshallian Externalities and the Emergence and Spatial Stability of Technological Enclaves*, Unpublished manuscript.

DeBresson, C. (1996), *Economic Interdependence and Innovative Activity*, Edward Elgar, Cheltenham.

Dicken, P. (1992), *Global Shift: The Internalization of Economic Activity*, second edition, Paul Chapman, London.

Dicken, P. (1994), 'Global-Local Tensions: Firms and States in the Global Space-Economy', *Economic Geography*, vol. 70, pp. 101-128.

Dierickx, I. and Cool, K. (1989), 'Asset Stock Accumulation and Sustainability of Competitive Advantage', *Management Science*, vol. 35, pp. 1504-1513.

Dosi, G. (1988), 'Institutions and Markets in a Dynamic World', *The Manchester School of Economics and Social Studies*, vol. 61, 2, pp. 119-146.

Edquist, C. (ed.) (1997), *Systems of Innovation: Technologies, Institutions, and Organisations*, Pinter Publishers, London.

Ernst, D. (1981), *Restructuring World Industry in a Period of Crisis – The Role of Innovation: An Analysis of Recent Developments in the Semi-conductor Industry*, UNCTAD, Geneva

Ernst, D. and Guerriei, P. (1997), *International Production Networks and Changing Trade Patterns In East Asia: The Case of The Electronics Industry*, Danish Research Unit for Industrial Dynamics (DRUID) Working Paper no. 97-7, Copenhagen.

Feis, H. (1934), *Europe – The World's Banker 1870-1914* (reprinted 1974 with a new introduction by the author and by C.P.Howland), Augustus M. Kelley, Clifton.

Fleenor, D. (1993), 'The Coming and Going of the Global Corporation', *Columbia Journal of World Business*, vol. 28, 4 , pp. 6-16.

Ford, H. (in collaboration with Samuel Crowther) (1922), *My Life and Work*, William Heinemann Ltd, London.

Foss, N.J. (1996), 'Higher-Order Industrial Capabilities and Competitive Advantage', *Journal of Industry Studies*, vol. 3, 1, pp. 1-20.

Foss, N.J. (1997), 'Resources and Strategy: A Brief Overview of Themes and Contributions', in N.J. Foss (ed.), *Resources, Firms and Strategies*, Oxford University Press, Oxford, pp. 3-18.

Freeman C. (1982), *The Economics of Industrial Innovation*, second edition, Pinter Publishers, London.

Freeman, C. (1991), 'Networks of Innovators: A Synthesis of Research Issues', *Research Policy*, vol. 20, 5, pp. 5-24.

Freeman, C. (1995), 'The National System of Innovation in Historical Perspective', *Cambridge Journal of Economics*, vol. 19, pp. 5-24.

Friis, P. and Maskell, P. (1980), *Om Alfred Weber og hans lokaliseringsteori*, Handelshøjskolen i København. Institut for Trafik-, Turist og Beliggenhedsforskning. Publ.6/80. [On Alfred Weber and his Theory on Location], Copenhagen Business School, Copenhagen.

Gambetta, P. (1988), *Trust: Making and Breaking Co-operative Relations*, Basil Blackwell, Oxford.

Gertler, M.S. (1995), 'Manufacturing Culture: The Spatial Construction of Capital', paper presented at the annual conference of the Institute of British Geographers, Newcastle Upon Tyne, January 3-6.

Gibbons, M., Limoges, C., Nowotny, H., Schwartzman, S., Scott, P. and Troiw, M. (1994), *The New Production of Knowledge*, Sage, London.

Gordon, D.M., (1995) 'The Global Economy: New Edifice or Crumbling Foundations?' *New Left Review*, no. 168, pp. 14-64.

Gregory, D. (1982), 'Progress Past and Present', *Progress in Human Geography*, vol. 6, pp. 115-119.

Griliches, Z. (1994), 'Productivity, R&D, and the Data Constraint', *American Economic Review*, vol. 84, pp. 1-23.

Grossman, G.M. and Helpman, E. (1993), *Endogenous Innovation in the Theory of Growth*, National Bureau of Economic Research (NBER), working paper no. 4527, Cambridge MA.

Hagedoorn, J. and Schakenraad, J. (1992), 'Leading Companies and Networks of Strategic Alliances in Information Technologies', *Research Policy*, vol. 21, pp. 163-190.

Håkansson, H. (ed.) (1987), *Industrial Technology Development: A Network Approach*, Croom Helm, London.

Håkansson, H. (1989), *Corporate Technological Behaviour: Co-operation and Networks*, Routledge, London.

Hallén, L. and Johanson, J. (1985), 'Industrial Marketing Strategies and Different National Environments', *Journal of Business Research*, vol. 13, pp. 495-509.

Hamel, G. and Prahalad, C.K. (1994), *Competing for the Future*, Harvard Business School Press, Boston.

Hatchuel, A. and Weil, B. (1995), *Experts in Organizations: A Knowledge-Based Perspective on Organizational Change*, Walter de Gruyter, Berlin.

Hatsopoulos, G.N., Krugman, P.R. and Summers, L.H. (1988), 'U.S. Competitiveness: Beyond the Trade Deficit', *Science*, vol. 241, pp. 299-307.

Hatzichronoglou, T. (1996), *Globalisation and Competitiveness: Relevant Indicators*, Organisation for Economic Co-operation and Development (OECD), Directorate for Science, Technology and Industry, Working Papers, IV, 16, Paris.

Holstein, W.J. (with S. Reed, J. Kapstein, T. Vogel and J. Weber) (1990) 'The Stateless Corporation' *Business Week*, May 14, pp. 52-59.

Hounshell, D.A. (1984), *From the American System to Mass Production 1800-1932*, John Hopkins University Press, Baltimore.

Hu, Y.-S. (1992), 'Global or Stateless Corporations Are National Firms with International Operations', *California Management Review*, vol. 34, 2, pp. 107-126.

Jaffe, A.B., Trajtenberg, M. and Henderson, R. (1993), 'Geographic Localization of Knowledge Spillovers as Evidence of Patent Citations', *Quarterly Journal of Economics*, vol. 63, pp. 577-598.

Jewkes, J. (1933), 'Theory of Location of Industries, by A. Weber', *Economic Journal*, vol. 43, pp. 506-507.

Kennedy, P. (1993), 'Preparing for the 21st Century: Winners and Losers', *The New York Review of Books*, February 11, pp. 32-44.

Lorenz, E. (1998), 'Trust, Contract, and Economic Cooperation', *Cambridge Journal of Economics, forthcoming*.

Loria, A. (1888), 'Intorno all'influenza della rendita fondiaria sulla distribuzione topografica delle industrie', *Atti della R.Accademia dei Lincei* Rediconti IV, 4:115-126, Roma.

Loria, A. (1908), 'Studi sulla topografia dell'industria', in *Verso la Giostizia Sociale* Roma.

Lundvall, B.-Å. (1985), *Product Innovation and User-Producer Interaction*, Industrial Development Research Series no. 31, AUC, Aalborg.

Lundvall, B.-Å. (1988), 'Innovation as an Interactive Process – from User-Producer Interaction to the National System of Innovation', in G. Dosi, C. Freeman, R. Nelson, G. Silverberg, and L. Soete (eds), *Technical Change and Economic Theory*, Pinter Publishers, London, pp. 349-369.

Lundvall, B.-Å. (1992a) 'Introduction', in Lundvall, B.-A., (ed.) *National Systems of Innovation: Towards a Theory of Innovation and Interactive Learning*, Pinter, London, pp. 1-19.

Lundvall, B.-Å. (ed.) (1992b), *National Systems of Innovation: Towards a Theory of Innovation and Interactive Learning*, Pinter Publishers, London.

Maddison, A. (1991), *Dynamic Forces in Capitalist Development*, Oxford University Press, Oxford.

Malecki, E.J. (1991), *Technology and Economic Development: The Dynamics of Local, Regional and National Change*, Longman, Harlow.

Malmberg, A. (1996), 'Industrial Geography: Agglomeration and Local Milieu', *Progress in Human Geography*, vol. 20, pp. 392-403.

Malmberg, A. and Maskell, P. (1997), 'Towards an Explanation of Regional Specialization and Industry Agglomeration', *European Planning Studies*, vol. 5, pp. 25-41.

Marshall, A. (1890), *Principles of Economics*, eighth edition 1920, reprinted 1990, Macmillan, London.

Marshall, A. (1919), *Industry and Trade: A Study of Industrial Technique and Business Organization, and of Their Influences on the Condition of Various Classes and Nations*, reprinted 1927, Macmillan, London.

Maskell, P. (1997), 'Learning in the Village Economy of Denmark: The Role of Institutions and Policy in Sustaining Competitiveness', in H.-J. Braczyk, P. Cooke, and M. Heidenreich (eds), *Regional Innovation Systems: The Role of Governance in a Globalized World*, Taylor and Francis, London, pp. 190-213.

Maskell, P. (1998), 'Low-tech Competitive Advantages and the Role of Proximity – The Case of the European Furniture Industry in General and the Danish Wooden Furniture Production in Particular', *European Urban and Regional Studies*, vol. 5, 2 (*forthcoming*).

Maskell, P. and Malmberg, A. (1995), *Localised Learning and Industrial Competitiveness*, Working Paper 80, Berkeley Roundtable on the International Economy (BRIE), available at Internet URL: HTTP://BRIE.BERKELEY.E-DU/BRIE/PUBS/WP/WP80.html.

Maskell, P., Eskelinen, H., Hannibalsson, I., Malmberg, A. and Vatne, E. (1998), *Competitiveness, Localised Learning and Regional Development: Possibilities for Prosperity in Open Economies*, Routledge, London.

Maunier, R. (1908), 'La Distribution Géographique des Industries', *Revue Internationale de Sociologie*, vol. 16, pp. 481-514.

Montgomery, C. A. (ed.) (1995), *Resource-Based and Evolutionary Theories of the Firm: Towards a Synthesis*, Kluwer, Boston.

Nelson, R.R. (ed.) (1993), *National Innovation Systems: A Comparative Analysis*, Oxford University Press, Oxford.

Nonaka, K. (1991), 'The Knowledge-Creating Company', *Harvard Business Review*, vol. 69, 6, pp. 96-104.

Patel, P. (1995), 'Localised Production of Technology for Global Markets', *Cambridge Journal of Economics*, vol. 19, pp. 141-153.

Patel, P. and Pavitt, K. (1991), 'Large Firms in the Production of the World's Technology: An Important Case of "Non-Globalisation"', *Journal of International Business Studies*, vol. 21, 1, pp. 1-21.

Penrose, E.T. (1959), *The Theory of the Growth of the Firm*, Oxford University Press, Oxford.

Polanyi, M. (1958), *Personal Knowledge: Towards a Post-critical Philosophy*, University of Chicago Press, Chicago; Routledge, London.

Polanyi, M. (1966, 1983), *The Tacit Dimension*, Routledge, London; Kegan Paul, New York.

Porter, M.E. (1990), *The Competitive Advantage of Nations*, Macmillan, London; Free Press, New York.

Prahalad, C.K. and Hamel, G. (1990), 'The Core Competence of the Corporation', *Harvard Business Review*, vol. 68, 3, pp. 79-91.

Predöhl, A. (1928), 'Theory of the Location and in its Relation to General Economics', *Journal of Political Economy*, vol. 36, pp. 371-390.

Rasmussen, H.K. (1996), *No Entry: Immigration Policy in Europe*, CBS Press, Copenhagen.

Reich, R.B. (1990), 'But Now We're Global', *Times Literary Supplement*, August 31-September 6, pp. 925-926.

Ricardo, D. (1817), *On the Principles of Political Economy and Taxation*, John Murray, London.

Richardson, G.B. (1997), *Economic Analysis, Public Policy and the Software Industry*, Danish Research Unit for Industrial Dynamics (DRUID) Working Paper no. 97-4.

Rivera-Batiz, L. and Romer, P.M. (1991), 'Economic Integration and Endogenous Growth', *Quarterly Journal of Economics*, vol. 106, pp. 531-555.

Rodrik, D. (1997), *Has Globalization Gone Too Far?* Institute for International Economics, Washington, DC.

Romer, P.M. (1990), 'Endogenous Technological Change', *Journal of Political Economy*, vol. 98 (Supplement), pp. 71-102.

Rosenberg, N. (1972), *Technology and American Economic Growth*, Harper and Row, New York.

Rumelt, R.P. (1984), 'Towards a Strategic Theory of the Firm', in R.B. Lamb (ed.), *Competitive Strategic Management*, Prentice-Hall, Englewood Cliffs, NJ, pp. 556-570.

Rumelt, R.P., Schendel, D. and Teece, D. J. (1991), 'Strategic Management and Economics', *Strategic Management Journal*, vol. 12 (Winter), pp. 5-29.

Sako, M. (1992), *Prices, Quality and Trust: Inter-firm Relations in Britain and Japan,* Cambridge University Press, Cambridge.

Sala-i-Martin, X. (1990). *Lecture Notes on Economic Growth (I): Introduction to the Literature and Neoclassic Models*, National Bureau of Economic Research (NBER), working paper no. 3563, Cambridge Mass.

Scott, B.R. and Lodge, G.C. (eds.) (1985), *US Competitiveness in the World Economy*, Harvard Business School Press, Boston.

Smith, A. (1776), *An Inquiry into the Nature and Causes of the Wealth of Nations* (reprinted in part as Penguin Classics, Harmondsworth 1979), W. Strahan and T. Cadell, London.

Spender, J.-C. (1994), 'The Geographies of Strategic Competence: Borrowing from Social and Educational Psychology to Sketch an Activity and Knowledge-based Theory of the Firm', Unpublished paper presented at the Prince Bertil Symposium: The Dynamic Firm, Stockholm.

Tobin, J. (1969), 'A General Equilibrium Approach to Monetary Theory', *Journal of Money, Credit and Banking*, vol. 1, pp. 15-29.

Torrens, R. (1808), *The Economists Refuted*, reprinted 1984, Augustus Kelley, Sydney.

Vickery, G. (with Claudio Casadio) (1996), 'The Globalisation of Investment and Trade', in J. de la Mothe and G. Pacquet (eds.), *Evolutionary Economics and the New International Political Economy*, Pinter, London, pp. 83-117.

von Hippel, E. (1994), 'Sticky Information and the Locus of Problem Solving: Implications for Innovation', *Management Science*, vol. 40, pp. 429-439.

Wade, R. (1996), 'Globalization and its Limits: Reports of the Death of the National Economy are Greatly Exaggerated', in S. Berger and R. Dore, *National Diversity and Global Capitalism*, Cornell University Press, Ithaca, pp. 60-88.

Weber, A. (1909), *Über den standort der industrien*, teil 1. (Translated by Carl Joachim Friedrich and published 1929 under the title: *Theory of the Location of Industries*, University of Chicago Press, Chicago).

Weber, A. (1923), 'Industrielle Standortslehre. Allgemeine und kapitalistische Theorie des Standortes', in *Grundriss der Sozialökonomik* (2nd. ed), Tübingen, pp. VI: 58-86.

Wernerfelt, B. (1984), 'A Resource-Based View of the Firm', *Strategic Management Journal*, vol. 5, pp. 171-180.

Wood, A. (1995), 'How Trade Hurt Unskilled Workers', *Journal of Economic Perspectives*, vol. 9, 3, pp. 57-80.

Worthington, R. (1993), 'Introduction: Science and Technology as a Global System', *Science, Technology, & Human Values*, vol. 18, pp. 176-185.

Young, A. (1994), 'Lessons from the East Asian NICS: A Contrarian View', *European Economic Review*, vol. 38, pp. 964-973.

Zander, U. and Kogut, B. (1995), 'Knowledge and the Speed of the Transfer and Imitation of Organisational Capabilities: An Empirical Test', *Organizational Science*, vol. 6, pp. 76-92.

Zysman, J. (1994), *National Roots of a 'Global' Economy*, Berkeley Roundtable on the International Economy (BRIE), University of California, Berkeley.

Zysman, J., Doherty, E., and Schwartz, A. (1996), *Tales from the 'Global' Economy: Cross-National Production Networks and the Re-organization of the European Economy*, WP 83, Berkeley Roundtable on the International Economy (BRIE), University of California, Berkeley.

3 Innovation and the City

Olivier Crevoisier

Introduction

What is the relationship between the city and economic development and, more precisely, between the city and technological change? Today, this very ordinary question gives rise to a host of questions. The traditional opposition between urban and rural areas no longer applies. In modern-day Europe, farming accounts for only a small share of labour and it is not easy to know any more what sets rural areas apart from urban areas. Lifestyles, economic activities, income levels – the elements which in the past meant that the rural could be considered as the traditional and somewhat backward system – have become considerably homogenised.

Today, one may well wonder whether the difference between urban and rural economic systems is still a question of nature (e.g. rural areas specialising in certain activities) or simply one of degree (e.g. a proportionately greater number of tertiary activities in urban areas). In Europe as a whole, owing to the development of communication links and infrastructures in general, even traditionally remote regions have access to the largest international centres in just a few hours – and sometimes in even less time than it takes to get around in a big city. Today, 'rural' inhabitants have access to the same jobs, engage in the same leisure activities (most often television!), and enjoy similar education and training opportunities as city dwellers. Does this mean that there are no longer any differences between these two types of areas? Has the city imposed once and for all its logic made up of 'networking of networks' throughout Europe? The question of the economic specificity of the urban must therefore be posed once again today. Accordingly, this chapter puts forward the *hypothesis* that the city still plays a special role in technological change and processes of innovation as compared with other areas. It may be noted straightaway that what is involved is exploring the long-term development processes of structural and relational change, not merely identifying the advantages in terms of flows – essentially of external economies of agglomeration – that firms derive from the city from the point of view of innovation and technological change.

The first part of the chapter shows how theories of development and technological change approach the urban. From the viewpoint of historians and for past centuries, matters seem clear. The city and development have historically been closely linked, in particular up to the Industrial Revolution. However, the analytical tools supplied by historians must be thoroughly rethought in order to take account of the present situation, for the transformations of the production system have imposed their logic more and more frequently on urban dynamics ever since the early 19th century.

Present literature on innovation and territorial development has today been clearly separated into two distinct currents, neither of which is satisfactory when it comes to the role of the city in innovation. On the one hand, thinking focuses on endogenous development (Becattini, 1990; Garofoli, 1992), industrial technopoles (Planque, 1985), innovative milieux (Maillat and Perrin, 1992; Maillat *et al.*, 1993), localised production systems (Colletis and Pecqueur, 1995; Courlet, 1994), learning regions (Asheim, 1996; Morgan, 1997), which virtually always ignore purely and simply both the urban and the non-urban dimension. Yet there is no gainsaying that the Italian industrial districts are based on an urban structure, that technopoles cannot be separated from the urban context and that the urban does in the final analysis play a role in the processes described. These approaches incorporate or ignore the urban in theories of development.

On the other hand, we find literature on metropolitanisation and global cities (Sassen, 1991; Scott, 1984; Veltz, 1996). These approaches describe accurately and theorise cogently the present development of cities – and in particular the largest cities – but do not in any event explain why these phenomena primarily take place in (large) cities. As a matter of fact, the city is postulated rather than explained. In such studies, the processes that are dealt with take place in a city. But does their location in a city play a specific role? The first part of the chapter discusses the ways in which these approaches incorporate or ignore non-metropolitan areas.

Why have these two currents remained isolated from each other until now? How can we conceive of development by implicitly postulating either that the urban plays no role or alternatively that it controls the entire process of development?

The question posed here is that of *the specific role of cities in technological innovation*. What role do cities play as compared with other areas that are only moderately or not at all urbanised? Since this is an old question which has already been amply covered with regard to previous historical periods, this study suggests that *very little effort has been put into explaining the role of contemporary cities in innovation and development*.

The second part of the chapter attempts to provide a more precise definition of these specific territorial characteristics. The *production system* develops and imposes its logic on territories based on innovation, the launching of new products on the market or the use of new technologies. The development of a city is grounded on a different logic. In what terms do we understand the (re)production of the city, in particular as compared with other areas?

The third part of the chapter puts forward the idea, based on Rémy and Voyé's (1992) work, that the city is characterised by a specific capacity to generate *interaction and learning sites* (ILSs) and foster their evolution. These are both constituent of the urban fabric and essential from the point of view of innovation in the production system. It will be shown that although the city is not necessarily the place where economic innovation occurs, it (re)constitutes a context which facilitates such innovation, thereby allowing considerable control over the evolution of the production system. In other words, the city is not necessarily the place most conducive to economic innovation; however, *it possesses a 'metacapacity' to develop ILSs* which generate resources and bring the economic actors who innovate into contact with each other. In the fourth part of the chapter, the specific role of cities is therefore characterised essentially by the various activities of bringing actors into contact with each other. It is suggested that this metacapacity is primarily manifested in two processes: *objectivisation of institutions* and *anchoring in the constructed*.

The Treatment of the Urban in Theories of Development and Innovation

Cities and Development: A Historical Viewpoint

For Braudel (1979), a number of characteristics are almost always attached to cities. Thus, there is no city without a somewhat complex division of labour. There is no city without a market, and there are no regional or national markets without cities. There are no cities without power which is both protective and coercive. Finally, there is no opening-up to the world and trade with faraway places without the city (p. 423). The city always dominates an area, a hinterland. Braudel compares the city to an electrical transformer which turns faraway high-voltage networks into low-voltage regional networks and vice versa. The city structures its hinterland based on distant profit opportunities and asserts the value of local production in faraway places. As far as this hinterland is concerned, the development of cities and economic development are to a large extent the same thing.

Such distinctions are not satisfactory when it comes to dealing with contemporary relations between cities and economic development. Indeed, in the Western economies, the division of labour between urban and other activities can no longer be clearly established, unless it be from a relative point of view: cities feature a greater concentration of tertiary activities and less farming than non-urban areas. Clearly, however, commercial firms that do business on a worldwide level can be localised today in areas with a relatively low degree of urbanisation. Thus, although these categories are essential for understanding the historical role of cities, today they absolutely must be completely rethought.

Yet the element which most clearly militates in favour of re-examining relations between cities and economic development is *technological change*, which we primarily apprehend through innovation (as defined by Schumpeter, 1935). As far as the Industrial Revolution is concerned, Braudel clearly shows that it did not take place in the main cities of the day but rather in Manchester, Birmingham, Sheffield, etc. – cities which were the products, not the incubators of the Industrial Revolution. Thus, cities do not have a corner in innovation, contrary to what some authors, such as Pred (1977), maintain. The dynamics of the production system sometimes gains the upper hand, imposing its logic on the urban structure, without however subjecting it. *It is precisely this interaction between the dynamics of the production system and the urban dynamics which is at the heart of our reflection.* Moreover, Mokyr poses the question of the relationship between technological change and the urban context and concludes plainly: 'it is possible to show that easy generalisations about the positive role of cities in technological progress are historically false' (Mokyr, 1995, p. 5). And further on, 'a more careful examination of the evidence reveals that notwithstanding a priori arguments, urbanisation has been neither necessary nor a sufficient condition for technological change' (p. 19). Thus, the question of the relationship between technological innovation and cities remains open.

Contemporary Approaches to Regional Development

As far as regional development is concerned, the past twenty years have been marked by theories on industrial districts (Becattini, 1990), technopoles and innovative milieux (Maillat and Perrin, 1992; Maillat *et al.*, 1993) and other territorial production systems (Colletis and Pecqueur, 1995; Courlet, 1994). All these different approaches highlight the role of networks of SMEs and the local insertion of large firms in the innovation process. In these approaches, *territory* is deemed to be an essential motivating force of development: the relations between the different local actors, be they merchants or non-merchants, make it possible to identify and mobilise local resources for innovation (Crevoisier, 1996). Territory is thus constituted by a quantity of relations (Storper, 1995) all of which are assets that can be mobilised in the economic process. The *innovative milieu* is deemed to be a localised set of interdependent actors who on the one hand jointly develop their know-how with a shared view of innovation and on the other hand develop between themselves the rules of competition/cooperation which renew the territory (Crevoisier, 1993).

In the literature, however, these networks between local actors are almost never explicitly situated in an urban or non-urban setting. *The underlying conception of territory is fundamentally relational, but it virtually always leaves aside the distinction between cities and rural areas.* In an exception to this rule, Courlet and Pecqueur (1992) show how the industrial districts in the Rhône-Alps region have tended over the past twenty years either to decline or to turn towards new forms which make much more intensive use of research and training centres. Moreover, these traditional industrial districts coexist with more recent *technological districts* which mobilise urban resources much more intensively than industrial districts.

The virtual absence of commentaries on the urban in current theories of territorial development can give rise to two interpretations. The first possibility is that the dynamics of the production system and technological change does not rely on the urban. Thus, the urban/rural or urban/non-urban distinction is not a discriminant from this point of view. The second possibility is that the relationship between innovation and the city is less direct than was previously thought and that research must be pursued along these lines.[1] We shall focus on the latter possibility.

Current Theories of Metropolitan Development

Another body of work focuses on the current development of metropoles. Empirically speaking, it can be seen that the largest cities in the Western world tend to grow rapidly and to account for an ever-increasing share of high value-added activities. For example, Veltz (1996) notes that in France between 1982 and 1990, population declined in the most sparsely populated areas, whereas the Ile-de-France area accounted for more than half (370,000) of all jobs created nationwide (700,000). Moreover, various studies consider that the development of metropoles compensates for the possible problems related to economic globalisation: the establishment of world production and distribution networks is accompanied by the management and supervision of economic activities which are supposedly the exclusive right of 'global cities' (Sassen, 1991). In this view, these metropoles have generated exceptional capacities in these management and supervision activities and have therefore become centres for innovation in the field of services provided to companies and tied to the supervision and management of economic activities. This vision is interesting insofar as it does not speak of technological change and innovation in the traditional sense, i.e. as tied to industry, but rather puts forward the argument that they are tied to new types of services in which metropoles have specialised. To sum up, the large cities generate new forms of domination over their 'hinterlands' which correspond to the management and service activities required by globally competing firms.

A good deal of research focusing more closely on technological innovation has been done on production systems located in large cities. For example, Suarez-Villa and Rama (1996) show that the electronics sector in Madrid has survived and developed thanks to relations between R&D centres and companies in the city.

> The intrametropolitan clustering of some firms may, for example, facilitate the outsourcing possibilities, contacts and information that would not be readily available with dispersion. Locating in a central area of the metropolitan region, rather than in periphery, may help to create and support networks of co-production or subcontracting that can be vital to R&D activities, through the resource savings that they provide (Suarez-Villa and Rama, 1996, p. 1156).

Their findings clearly show that the fact that these activities are located inside Madrid favours innovative capacity. Yet is location in a city a prerequisite for this?

Traditional approaches in economics have explained the city by external economies of urbanisation and localisation. However, as far as technological change and innovation are concerned, such approaches remain unsatisfactory. When Marshall, the inventor of the concept, explained external economies, he illustrated them by the industrial districts of England in the late 19th century – yet these districts did not necessarily take the form of a city. Not every externality is a component of a city, even through the city is an area which can generate externalities. In this connection, studies on metropolitanisation have likewise failed to bring out clearly the role of urban specificities in innovation.

For Camagni (1995), *proximity and agglomeration do not necessarily coincide*. Authors like Scott (1988), however, associate proximity with agglomeration when speaking about California. This assimilation might be a consequence of differences between European and North American circumstances. In North America, outside major urban agglomerations, few regions feature the same population density and activity characteristics that are found in corresponding areas in Europe (e.g. in Germany, Northern Italy, Switzerland, Benelux).

Thus, explanations of technological change in metropoles highlight the advantages of proximity but fail to specify exactly why and how proximity would serve innovating actors in the city specifically. The following sections aim to clarify this missing link in explanations of the location of innovative activities.

The Territorial Characteristics of the City

How can the distinction between urban and non-urban areas be reintroduced in theories of innovation? Here, the best method consists of starting off on the basis of both the material and social characteristics of the city which unmistakably differentiate it from non-urban areas. Using this as a starting point, we will explain the relations between urban dynamics and innovation based on the notion of *interaction and learning sites* (ILSs). These sites are *simultaneously* generated, developed and destroyed by the city and they participate in the process of innovation, in particular by bringing economic actors into contact with each other and enabling the transfer of know-how. Thus, the relationship between city and innovation is not a direct one. Instead, first, the specific nature of the city lies in its ability to *generate ILSs*. Second, ILSs produce and maintain the *resources for innovation*.

A Territorial Approach to the City

Rémy and Voyé (1992) define the city as a social and material unit. In this context, we may understand 'social unit' as a set of actors related by institutions as defined below. The city is:

> ... a social unit, made up of a set of interrelated social functions which, through the convergence of products and information, plays a special role in exchanges, be they material or other; a material unit, characterised by a certain density as well as a continuity of the constructed, within which one observes a series of constituent oppositions (centre/suburbs, private/public areas, etc.). When viewed from this morphological perspective, the city derives its specificity from the fact that it is neither the place where a specific function is exercised (as is the case for a house, a school, a hospital, a firm) nor the place where these specific functions are juxtaposed, but rather the place which interrelates these various functions, through the spatial relationship (Rémy and Voyé 1992, p. 8).

This definition views the city first of all as an area constructed in a continuous fashion, which clearly distinguishes it from other kinds of territories. Second, it describes the city as a place which interrelates these different functions. The city appears as a setting or more exactly a 'meta-place' of interaction, with both a social component (made of actors interrelated by and generating institutions) and a material (constructed) component. Thus, it is the articulation and the superimposition of these different functions in an area which characterises *the city as distinct from non-urban areas*.

Relations between the City and Innovation

Interaction and Learning Sites   How then can these characteristics of the city be linked to innovation? As a working hypothesis, I propose that these two components, the social (the institutional) and the material (the constructed), can be joined by using the notion of *interaction and learning sites* (Figure 3.1). In concrete terms, these ILSs may be research and training centres, professional associations (lobbies, trade unions, sectoral bodies, employers' associations), trade fairs, centres for technology transfer, specialised media, and sometimes libraries or museums. Sometimes, they may be mere meeting places: for example the hotel where the heads and the key decision makers of firms in a specific business meet in an informal setting.

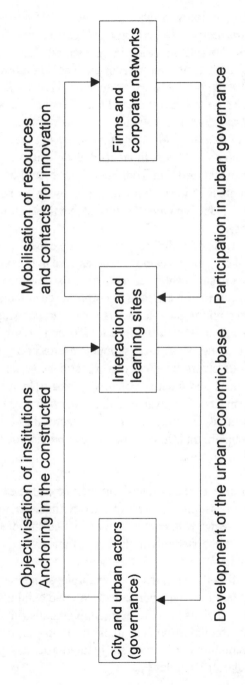

Figure 3.1 The interactions between urban and innovative dynamics through ILSs

The Role of ILSs in Innovation: Establishment of Contacts and Creation of Innovative Resources   The concept of ILSs injects an essential element to explanations of the role of cities in innovation. Innovating firms are not necessarily located in a city but parts of the innovation process almost always pass through the city at one stage or another. The city is not necessarily the place most conducive to innovation as such. However, *the function cities serve in innovation processes is that they develop various types of ILSs where some of the resources required in innovation processes develop and come into contact with each other.* These thoughts dovetail with those of Keeble (1993) and, to a lesser extent, those of Tödtling (1990, 1995), which show that firms located in rural areas or in small towns are innovative – sometimes even more so than firms located in big cities – provided that they enjoy satisfactory access to urban centres.

Different ILSs play different roles. They can be the places where new types of know-how are developed, such as research centres. Naturally, the people who work in such centres do more than just develop technical skills: they also establish contacts with colleagues, corporate customers, and others. Trade fairs play a very important role in this connection as well. They give firms operating in a given (specialised) sector an opportunity to see what their competitors are doing, what their suppliers are offering, and to meet a very broad range of customers. They offer an exceptional occasion to make contacts, come up with ideas and identify potential partners. In sum, ILSs supply two essential components of innovative milieux (Crevoisier, 1993; Maillat *et al.*, 1991): new resources in terms of know-how, as well as essential contacts and interactions to ensure that these resources can be incorporated and developed in a process of innovation.

The Materialisation of ILSs in the City   As far as the city is concerned, ILSs reinforce the two essential aspects of the city: its capacity to bring different functions into contact with each other, and its constructed component. They may be materialised differently in the urban fabric:

- permanently or temporarily: for example, some major cities have large trade fair facilities, with permanent staff and buildings; conversely, some trade fairs, such as the ones in the Italian industrial districts, are held in temporary facilities (such as tents) or multi-purpose facilities (such as concert halls) and are run by staff temporarily assigned to such events;
- on a regular or irregular basis;

• for a single purpose or for various purposes: for example, some research centres may limit themselves to specialised research, while others also handle training, technology transfers and perhaps lobbying as well.

One essential element of ILSs is that they constitute economic activities in themselves, even to the extent that they represent the city's most important activities. Numerous studies demonstrate the economic significance of universities, research centres (e.g. Felsenstein, 1996), and international trade fair activities. Let us imagine cities without higher educational facilities, without research centres, without trade fairs, without the different occupational associations. In most cases, their main role would be limited to administration and the retail trade.

The Genesis of ILSs: Urban Dynamics and Innovative Dynamics The ILSs are thus situated at the crossroads of two dynamics. On the one hand, they are solicited, encouraged, funded, run or used by firms. On the other hand, they are backed, funded, lodged, run and attracted by public or institutional actors or by the other ILSs already set up in a given city. This dual role enables us to better characterise the role of cities in innovation and technological change. It makes us ask: How can such sites be generated and sometimes reoriented and developed? We argue that this process can essentially be described through two *functions* performed by cities, namely 'objectivatisation of institutions' and 'anchoring in the constructed' (Corolleur *et al.*, 1996).

Objectivisation of Institutions The city, as a social system, is a system that consists of a variety of actors and their relations. The actors naturally include firms and inhabitants, but also specifically urban organisations such as training and research institutes, occupational and trade union associations, political and administrative authorities. Contemporary institutional economists focus on the relations between social actors. Consequently, *institutions* can be defined as rules which concern relations between actors: formal procedures (e.g., contracts, regulations, laws, etc.) or informal procedures (e.g., modalities of competition, cooperation, division of labour, trust or mistrust). They are 'the rules of the game in society or … the humanly devised constraints that shape human interaction' (North, 1990, quoted by Hodgson, 1998). This definition can be applied to both individual and collective or organised actors. *Institutionalisation* is the process of developing rules (i.e., the establishment of a regularity of behaviour of a longer duration).

Institutions are at the same time embedded in individual economic actors and shared by a certain number of them. In the line of this idea, we suggest that territories may be characterised by their institutions, following different degrees of *objectivisation*, that is to say the degree of autonomy that a specific institution has relative to given individuals. On the one hand, the least urbanised milieux are governed by institutions based on interpersonal or inter-group relations. The institutions are strongly related to individuals up to the point at which it is difficult and meaningless to differentiate relations between individuals and institutions. On the other hand, the urban context seems to be particularly conducive to a process of *objectivisation*, to a disconnection between individuals and institutions. Institutions are more 'external' to individuals; they get a stronger autonomy towards individuals who created or animate them. Of course, this process occurs very often in parallel with a process of formalisation of institutions, but it cannot be reduced to it. The urban context facilitates the passage from relations strongly embedded in particular individuals or groups to more independent structure. For example, a body which starts out as a group of industrialists who meet once a week informally in a relatively non-urbanised setting is rapidly attached to an already existing formal association endowed with an address, a budget, an organisational structure, and a staff, in a more urbanised context.

The urban context provides ready-to-use institutions which allow the passage between interindividual action to collective and organised action. A training programme set up by a group of firms in a non-urbanised setting is subsequently entrusted to an existing organisation in the city. A regional trade fair, set up by firms, is likely to be run subsequently by people specialised in organising this type of event in a city. A survey by Corolleur *et al.* (1996) showed that projects originating in different professional milieux ended up taking the form of new organisations (such as training centres for given industries or professional associations) in cities. This is how the majority of training and research institutions, professional associations, and other organisations come into being. First, there is an informal project backed by a few individuals. These individuals seek to interest more and more actors so as to give concrete form to their project. Inasmuch as the urban context is receptive, these projects take the form of organised societies with an organisational structure, stable financial resources, and so on. The objectivisation of institutions essentially takes place via the interaction between the various urban actors and pre-existing associations or institutions. Thus, it is the *multiplicity of urban actors* and their capacity for interaction which facilitate the integration or consolidation of new actors.

Accordingly, the city, through its capacity to objectivise institutions, facilitates the passage from occasional actions based on voluntary participation and a militia system to organised, continued actions carried out by professionals. From the point of view of its impact on capacities for innovation, the city systematises interactions and learning by structuring the framework and stabilising its funding. These organisations and institutions can in turn become a factor of attraction for new organisations thanks to the interplay of economies of scale, reputation and image effects, and above all by junction and complementarity effects which are particularly strong in view of the fact that the urban fabric has numerous active actors. Along the same lines, this appearance or consolidation of urban actors thus is one of the key characteristics of 'the urban'.

The urban context is particularly conducive to this process of objectivisation of institutions. At the same time, it would appear that this is linked to an *anchoring in the constructed*, i.e. to a location in a city, often in a building, and not on an *ad hoc* basis.

Anchoring in the Constructed   The objectivisation of institutions may be projected onto the constructed. The city, as a material unit characterised by a certain density and continuity of the constructed area (Rémy and Voyé, 1992), plays a special role in the (re)production of capacities for innovation by centralising this activity or even giving it a material infrastructure on an *ad hoc* basis.

This anchoring facilitates the perpetuation of institutions and fosters the passage from diffuse actions to actions which are organised and perhaps centralised. Moreover, because buildings are a characteristic element of cities, they in turn constitute an essential characteristic of the city itself. The development of new buildings and new sites also facilitates the emergence of new organisations by giving them a constructed basis. For example, economies of scale have appeared for buildings in training, research, trade fairs, ocupational associations or public bodies. Physical proximity within a given city can also favour junction and complementarity effects with other ILSs. Often, cities become involved in the setting-up of the ILSs by making premises or land available. The emphasis on these two functions is close to the conception of Rémy and Voyé (1992, p. 8), for whom, as noted above, the city derives its specificity from the fact that it is a site of interaction, with both a social (institutional) and a material (constructed) component, which integrates different functions through the spatial relationship.

The Blocked City   All cities do not fulfil these two functions in the same way. In an ideal case, relations between actors are such that neither the objectivisation of institutions nor the anchoring in the constructed poses a problem. The existence and the effective implementation of these processes are always subject to the interplay of actors, to the form of government of a city, to the relations between different bodies, between public and private sectors, in short, to what is now called *governance* (Stoker, 1998). The competitive and/or cooperative relations between actors influence urban dynamics and the dynamics of innovation.

Cities which have not managed to ensure enough collaboration between local actors to facilitate the processes of the objectivisation of institutions and the anchoring in the constructed can be called *blocked cities*. There are many different types of blockages: separation of the political and economic spheres, competition between the various organisations, competitition between existing organisations and projects under way, indifference, blockages in processes involving the regeneration of urban sites and the reallocation of land and buildings. Such blockages can have various effects, such as weakening of innovative dynamics and the positive impact of the city, or decisions to locate ILSs in other cities.

Conclusions: Rethink the Urban in Technological Change

This chapter has focused on a very specific aspect of cities, namely, their role in innovation and technological change. This is an important issue from several points of view.

First, we suggest in this paper that there exists a division of labour, hence complementarity, between cities and less or non-urbanised areas. Technological change and innovation do not occur more easily or more frequently in cities than elsewhere, at least in countries where infrastructures and public services are well developed throughout the territory. Nevertheless, this does not mean that cities do not have a specific role in innovation processes. Cities seems to provide *interaction and learning sites* that provide the resources for innovation and allow connections for innovation to occur. Consequently, innovative firms do not need to be located in cities, but they need to have good connections with ILSs. In terms of regional or urban planning, this idea may be of considerable importance.

Second, we underlined the importance of the process of generation of ILSs. We may distinguish two classical cases: exogenous and endogenous development. On one hand a certain number of cities fight each other to attract such ILSs. There exist a fierce competition at national and international level in order to get the most important research centers, universities, international fairs, professional bodies, media, libraries, etc. All cities are not able to get a part of these. Only the best endowed in terms of public services, geographical location, finances and overall political support may receive some types of ILSs. In this case, representing the exogenous development process, ILSs provide standardised services, not specifically suited to the needs or characteristics of local firms. One the other hand, according to the logic of endogenous development, cities can support their existing innovative milieux and industrial districts and extend the innovative dynamics of firms located in their hinterland by supporting the development of sites active in the same specialisation as local firms. Depending on how active cities are in this regard and on how local governance is developing, they can be engines or blockages for development as far as the innovative dynamics of their own and their surrounding regions are concerned. This is the logic of the specialised city as opposed to the metropolitan city (Cattan and Saint-Julien, 1996).

Acknowledgements

This chapter takes up part of the problematic and concepts of a study (Corolleur *et al.* 1996) carried out within the framework of the European Research Group on Innovative Milieux (GREMI). As it was a group effort, I wish to thank my coauthors without whom the content of this paper would have been significantly different.

Note

1 This latter possibility is currently being explored within the framework of the European Research Group on Innovative Milieux (GREMI).

References

Asheim, B. (1996), 'Industrial Districts as "Learning Regions": A Condition for Prosperity?', *European Planning Studies*, vol. 4, pp. 379-400.
Becattini, G. (1990), 'The Marshallian Industrial District as a Socio-economic Notion', in F. Pyke, G. Becattini and W. Sengenberger (eds), *Industrial*

Districts and Inter-firm Cooperation in Italy, International Labour Office, Geneva, pp.37-51.

Braudel, F. (1979), *Civilisation Matérielle, Économie et Capitalisme: XVè au XVIIIè Siècle*, Armand Colin, Paris.

Camagni, R. (1995), 'Global Networks and Local Milieu: Towards a Theory of Economic Space', in S. Conti, E.J. Malecki and P. Oinas (eds), *The Industrial Enterprise and Its Environment: Spatial Perspectives*, Avebury, Aldershot, pp.195-215.

Cattan, N. and Saint-Julien, T. (1996), 'Le réseau intégré des villes en Europe occidentale: à l'articulation de plusieurs modèles urbains', Colloque de l'ASRDLF Régions et villes dans l'Europe de l'an 2000, Berlin.

Colletis, G. and Pecqueur, B. (1995), 'Politiques technologiques locales et création de ressources spécifiques', in A. Rallet and A. Torre (eds), *Économie Industrielle et Économie Spatiale*, Economica, Paris, pp. 445-463.

Corolleur, F., Boulianne, L.-M., Crevoisier, O. and Decoutère, St. (1996), 'Ville et innovation: le cas de trois villes de Suisse occidentale', Working Papers 9601, Institut de Recherches Économiques et Régionales, Neuchâtel.

Courlet, C. and Soulage, B. (eds) (1994), *Industrie, Territoires et Politiques Publiques*, L'Harmattan, Paris.

Courlet, C. and Pecqueur, B. (1992), 'Les systèmes industriels localisés en France: un nouveau modèle de développement', in G. Benko and A. Lipietz (eds), *Les Régions qui Gagnent – Districts et Réseaux: Les Nouveaux Paradigmes de la Géographie Économique*, Presses Universitaires de France, Paris, pp. 81-102.

Crevoisier, O. (1993), *Industrie et Régions: Les Milieux Innovateurs de l'Arc Jurassien*, EDES, Neuchâtel.

Crevoisier, O. (1996), 'Proximity and Territory versus Space in Regional Science', *Environment and Planning A*, vol. 28, pp. 1683-1697.

Felsenstein, D. (1996), 'The University in the Metropolitan Arena: Impacts and Public Policy Implications', *Urban Studies*, vol. 33, pp. 1565-1580.

Garofoli, G. (1992), 'Les Systèmes Industriels de Petites Entreprises: Un Cas Paradigmatique de Développement Endogène', in G. Benko and A. Lipietz (eds), *Les Régions qui Gagnent – Districts et Réseaux: Les Nouveaux Paradigmes de la Géographie Économique*, Presses Universitaires de France, Paris, pp. 57-80

Hodgson, G. (1998), 'The Approach of Institutional Economics', *Journal of Economic Literature*, vol. 34, pp. 166-192.

Keeble, D. (1993), 'Small Firm Creation, Innovation and Growth and the Urban-Rural Shift', in J. Curran and D. Storey (eds), *Small Firms in Urban and Rural Locations*, Routledge, London, pp. 55-78.

Maillat, D. (1995), 'Territorial Dynamic, Innovative Milieus and Regional Policy', *Entrepreneurship and Regional Development*, vol. 7, pp. 157-165.

Maillat, D., Crevoisier, O. and Lecoq, B. (1991), 'Réseau d'innovation et dynamique territoriale. Un essai de typologie', *Revue d'Économie Régionale et Urbaine*, no. 3-4, pp. 407-432.

Maillat, D. and Perrin, J.-C. (eds) (1992), *Entreprises Innovatrices et Développement Territorial*, GREMI, EDES, Neuchâtel.

Maillat, D., Quévit, M. and Senn, L. (eds) (1993), *Réseaux d'Innovation et Milieux Innovateurs : Un Pari pour le Développement Régional*, GREMI, EDES, Neuchâtel.

Mokyr, J. (1995) 'Urbanization, Technological Progress, and Economic History', in H. Giersch (ed.), *Urban Agglomeration and Economic Growth*, Springer, Heidelberg, pp. 3-38

Morgan, K. (1997), 'The Learning Region: Institutions, Innovation and Regional Renewal', *Regional Studies*, vol. 31, pp. 491-503.

Pecqueur, B. (1987), 'Tissu économique local et systèmes industriels résiliaires', *Revue d'Économie Régionale et Urbaine*, no. 3, pp. 369-378.

Planque, B. (1985), 'Le développement des activités à haute technologie et ses répercussions spatiales', *Revue d'Économie Régionale et Urbaine*, no. 5, pp. 911-941.

Pred, A. (1977), *City Systems in Advanced Economies*, Hutchinson, London.

Rémy, J. and Voyé, L. (1992), *La Ville: Vers une Nouvelle Définition?*, L'Harmattan, Paris.

Sassen, S. (1991), *The Global City: New York, London, Tokyo*, Princeton University Press, Princeton, NJ.

Schumpeter, J. (1935), *Théorie de l'Évolution Économique*, Dalloz, Paris.

Scott, A.J. (1988), *Metropolis: From the Division of Labor to Urban Form*, University of California Press, Berkeley.

Stoker, G. (1998), 'Cinq propositions pour une théorie de la gouvernance', *Revue Internationale des Sciences Sociales*, no.155, pp. 19-30.

Storper, M. (1995), 'L'économie de la région: les relations comme actifs économiques', Colloque de l'ASRDLF, Toulouse.

Suarez-Villa, L. and Rama, R. (1996), 'Outsourcing, R&D and Pattern of Intra-metropolitan Location: The Electronics Industries of Madrid', *Urban Studies*, vol. 33, pp. 1155-1197.

Tödtling, F. (1995), 'The Innovation Process and Local Environment', in S. Conti, E.J. Malecki, and P. Oinas (eds), *The Industrial Enterprise and its Environment: Spatial Perspectives*, Avebury, Aldershot, pp. 171-193.

Tödtling, F. (1990), *Räumlich Differenzierung betrieblicher Innovation*, Sigma, Berlin.

Veltz, P. (1996), *Mondialisation Villes et Territoires: L'économie d'archipel*, Presses Universitaires de France, Paris.

4 Local Product Development Trajectories: Engineering Establishments in Three Contrasting Regions

Neil Alderman

Introduction

Over the past 15 to 20 years, researchers have expended considerable effort studying the regional dimensions to technological change. This work, together with that of other scholars, has repeatedly identified significant regional differences in levels and rates of innovation in British industry, particularly in relation to product innovation (Thwaites *et al.*, 1981, 1982; Alderman, 1986; Alderman *et al.*, 1988; Alderman, 1994; Harris, 1988; Oakey *et al.*, 1988; Northcott and Rogers, 1984; Phelps, 1995). Explanations for the regional variations in innovation activity reported in the literature have tended to be limited to the identification of broad factors that are associated with such differences, such as industrial sector, ownership status, size and so forth (see for example, Tödtling, 1990). These 'explanations' offer little by way of an understanding of technological change, because they address the innovation event, rather than the underlying processes generating the event (Maillat, 1991).

An understanding of the process of technological change has been sought through the study of the micro-level processes by which firms and establishments bring about product development, both internally within the firm or manufacturing unit (e.g. Johne and Snelson, 1990; Braiden *et al.*, 1993; Alderman *et al.*, 1996) and in terms of its relations with its immediate local environment (e.g. MacPherson, 1992; Grotz and Braun, 1993; Sternberg, 1997). However, while this micro-level analysis offers considerable detail of the

79

processes engaged in by individual establishments, a robust explanation of aggregate regional patterns remains elusive.

This chapter reports on some results from a recent study that attempted to bridge the gap between the two levels of analysis by investigating the technological development of specific engineering products in matched establishments in three contrasting regions (Alderman and Thwaites, 1997). The research set out to assess the influence of accumulated experience and 'path-dependency' in the product development trajectories of the matched establishments and the extent to which this is determined by the particular conditions pertaining at the local level, both within the establishment and within its immediate environment. The principal hypothesis was that different conditions in regions with radically differing histories and industrial and organisational structures would give rise to different types of technological trajectory at the local level.

The chapter begins by outlining some of the conceptual background behind this thinking. The subsequent sections describe the methodology of the research project and present some results pertaining to the nature of the technological trajectories observed. The analysis focuses on the extent to which establishments in the different regions engaged with their local environment (local connections) and on the extent to which national or international networks of customers, suppliers and other agents (non-local connections) were crucial in the process of developing their products. Finally, some conclusions regarding the importance of local connections in the context of the UK engineering industry are drawn.

Background

It is widely accepted that technological innovation is as much a social process as it is about technical solutions to clearly specified problems. As such it does not take place on the head of a pin, but in real places that possess an environment that impinges on the establishment and influences its management. This environment reflects not just local factors, but also the juxtaposition of the local with both national and international economic conditions and the particular corporate environment of the establishment. The interaction between these (general) conditions and local conditions produces particular outcomes in different localities (Massey, 1984) and thus gives rise to different aggregate patterns of innovation performance.

This social process of technological innovation involves interaction between individuals within the system, both internally within the firm and between members of the firm and outside agencies and establishments. The network school of research (Håkansson, 1987; Biemans, 1992) seeks to explain the innovation process in terms of the network relations between these various actors. Networking principles are seen to apply both within and between organisations (Cooke and Morgan, 1993). Internally, networking occurs through new forms of product development practice (e.g. Takeuchi and Nonaka, 1986; Wheelwright and Clark, 1992) and externally through relations between the producer, its customers, suppliers and other agents (Lundvall, 1988; von Hippel, 1988). Under this model the trajectory of development of a product is not solely an outcome of an internally-focused development process, but also depends on co-operation between different actors in the system (Asheim and Cooke, Chapter 6, *this volume*).

The work of the GREMI school argues forcibly that such micro-level networks are not independent of their environment and that territorial factors have a fundamental role to play in the innovation process (Perrin, 1991). Successful regions are deemed to constitute an innovative 'milieu' (Maillat, 1990), which possesses key factors for innovation (Aydalot and Keeble, 1988), such as access to venture capital, local contact networks, and skilled labour. The local milieu is likely to reflect characteristics of the local business culture, which will influence local attitudes and approaches to technological change (Malecki, 1995). The precise mechanisms by which these factors are incorporated into the process of new product development at the micro scale, however, are not identified. The process of technological change within the establishment therefore essentially remains a 'black box' and policy makers tend to be left to attempt to 'lubricate' the technical process through financial or other assistance towards the development or adoption of new technologies, while failing to address the organisational and managerial structures that underpin this technical innovation activity.

The starting point for this research was a recognition that the development of products tends to follow an evolutionary pattern (Nelson and Winter, 1982). The individual manufacturing establishment will, to a greater or lesser extent, be subject to the pattern of its historical development, its inherited technological direction and skills base, patterns of managerial thinking, established contact networks and its organisational form. These provide the 'context conditions' for technological development within the establishment, so that the search for solutions to technical problems will be focused in such a way as to build on the existing technological base of the establishment (Dosi, 1988), much of which will be tacit, having derived from 'learning by doing'. Hence, 'most firms do (and are best at doing) what they have done in the past' (Sharp, 1990, p. 97). The result is a particular trajectory of technological development fashioned both by these firm- or establishment-specific conditions and by the prevailing technological paradigm (Dosi, 1982).

Technological change involves the development of a base of knowledge and the capability to transform that knowledge into new products and the procedures for manufacturing them. While some of this knowledge is written down in manuals and papers, much of it is tacit and resides with individuals in the employ of the firm or establishment (Dosi, 1988). This tacit knowledge forms an important component of the competitive advantage of the firm (Maskell, Chapter 2, *this volume*). This is especially true of traditional sectors where much of the accumulated experience of the establishment is often not captured in permanent media, but instead remains encapsulated in the experience and craftsmanship of key individuals. This learning also arises from different sources: learning by doing, using, searching, interacting, and from both internal and external activities (Malerba, 1992). Patterns of learning and the accumulated stock of knowledge will therefore be highly firm or establishment specific.

The notion of technological trajectories has been used to account for trends in the organisation of production systems (e.g. Piore, 1992), but there is a dearth of empirical evidence at the micro level to indicate the nature of technological trajectories in the products of individual establishments. Malerba (1992) argues that most product development is incremental and that the particular product trajectory will reflect the different learning patterns and sources of knowledge acquired by the firm. Similarly, Chesbrough (1994) has suggested that technological advance depends upon the idiosyncratic activities of firms – established routines and interpretation of heuristics – and Walsh (1996) has demonstrated that the design function within the firm exhibits a 'diffuseness' with the result that 'different structures have evolved which best meet the needs of each individual firm' (p. 525).

If one accepts the principles of trajectories and networking, it follows that product development trajectories at the local level will reflect, amongst other things, the internal historically determined pattern of development and knowledge accumulation of the establishment and the nature of its networking interactions with other actors in the system. Part of this interaction will take place with, and within the context of, the local environment of the establishment, particularly in relation to the transfer of tacit knowledge (see Maskell, Chapter 2, *this volume*), and this may be hypothesised to have an influence on the nature of the technological trajectory. Since local environments vary considerably, if local networking is important, one should expect to find systematic differences in trajectories in different locations.

Research Methodology

The principal objective of the research was to test whether there were systematic differences in the technological trajectories of products between regions that were known to differ in terms of their aggregate innovation patterns and in terms of their economic, social and general industrial characteristics. This required that, in order to control for extraneous factors, the study involve matched pairs of establishments between the chosen regions. The need to look in some detail at the product development activities and procedures of the establishments, in order to understand how product development actually took place, further dictated that the research pursue in-depth case studies.

Regions and Hypothesised Trajectories

The three UK regions selected for this study were the Northern region, the West Midlands and parts of the South East. The Northern region is characterised by an historical reliance on heavy engineering, ship building and coal, continuing structural problems of decline and its recent partial renaissance on the back of large-scale foreign inward investment, particularly in automotives and electronics. The West Midlands comprises the old industrial heartland founded on the iron industry and developing subsequently as the original home of the automotive and related industries, again with a strong engineering tradition. The South East is a region with a rather different engineering legacy and home to many of the newer industries, with a much more diversified economy and accessibility to the financial and commercial capital of the UK and both national and international markets.

The principal expectation concerning technological trajectories at the local level is that in the periphery (North), and perhaps to a lesser extent in the old industrial heartland (West Midlands), one should find a tendency towards mature products and trajectories that are characterised by incremental rather than radical innovation, slower rates of change and a focus on manufacturability and process (operational) considerations. In the (arguably) more dynamic environment of the South East one should find a greater preponderance of trajectories based on fast moving, young products featuring more frequent radical innovation and a focus on technical performance considerations. Stereotypes such as these would help to account for differences in regional innovation rates identified in previous research.

Selection of Establishments and the Matching Process

The requirements of the study created a complex research design, which involved attempting to match establishments in triplets. A target of 36 establishments, 12 in each region, was set. Establishments were matched as far as possible in terms of their size, ownership status, product characteristics (complexity)[1] and the nature of their markets. This degree of matching is rather more sophisticated than is usual in matched pair studies (Peck, 1985; O'Farrell and Hitchens, 1988). A refusal to participate could have resulted in the need to find two or even three replacements. For this reason, coupled with the increasing difficulty of finding companies with the time and resources to devote to detailed case study research, the Northern region case studies were completed first.[2] These were then matched against establishments in the other two regions.

It was necessary to obtain more case studies than eventually required in the Northern region owing to problems of matching.[3] Within the West Midlands, successful matches were obtained from Warwickshire, Staffordshire and Hereford and Worcestershire and the West Midlands county itself. In the South East, successful matches were located in the counties of Berkshire, Oxfordshire, Hertfordshire, Buckinghamshire and Essex, all sites being outside the M25.[4] Restricting the area of the South East to be included in the study obviously constrained the matching process. The result was a total of 33 pairwise matches between 38 establishments in the three regions.

The establishments were drawn from the mechanical, electrical and instrument engineering sectors, although the majority were in mechanical (including some in automotive components) engineering. Potential establishments were identified from the CURDS Diffusion Survey (Alderman, 1994) and from directory sources such as *Kompass* and *Key British Enterprises*, supplemented where necessary by local authority directories. The study targeted establishments of medium size (in accordance with the latest European size definitions), although one or two large ones were included. The establishments studied ranged in size from 55 to 700 employees, with a median of 170.

Data Sources

The primary method of data collection was through in-depth interview with technical and other staff within the establishment, but a wide variety of other sources were brought to bear on each case study. Many of the establishments had a multiple product portfolio and the studies focused on the development of the strategically most important product from the establishment's viewpoint.

The core of each case study comprised basic information about the establishment, its history, organisational structure, products and markets, and competitive drivers. It then dealt with issues surrounding the establishment's approach to product development, strategy, pace of change, technology monitoring, use of external sources of finance or technical expertise, relationships with customers and suppliers and the kinds of activities used to inform the product development process. Finally, the studies looked at a specific development project, focusing on its organisation, design strategy, supply items, manufacturing implications and the involvement of different functions in the process. Some general issues concerning the labour market and skills requirements for product development and the local business culture/environment were also addressed.

Characterising Technological Trajectories

The research required that a set of operational parameters be derived to enable individual trajectories to be characterised and compared. It was necessary to be able to compare different products rather than the same products from different manufacturers (cf. Gardiner, 1984); therefore, the definition of trajectories in terms of quantitative performance indicators, as is more usually the case (Georghiou *et al.*, 1986; Chesbrough, 1994), was not possible, since performance metrics would not be comparable. Instead, a series of qualitative indicators of the nature of the trajectory were derived to reflect how the product had developed over time and the nature of innovatory change, together with the establishment's strategic orientation towards technology. The indicators used are described below.

Pace   This reflects the rate of technological change in the product. The trajectories are categorised as *slow* if the product experiences fundamental technological innovation less frequently than every 10 years. *Fast* trajectories experience fundamental technological change on a regular basis (every five years or less), while *medium*-paced trajectories are defined as those where fundamental changes occur every five to ten years. In many cases this parameter reflects changes in technology that are external to the establishment, unless it happens to have been the originator of the change.

Design Status   This is based on Hollins and Pugh's (1990) notions of *static* and *dynamic* products. This parameter is primarily an indicator of internally generated changes to the technology as it reflects the degree and sophistication of design activity applied to the product within the establishment.[5] Static products are those where the design is not changing much or is unlikely to change. The emphasis is likely to be on serving a stable market with conservative customers, design for manufacture and cost reduction or value engineering, standardisation and parts rationalisation, and strictly incremental changes to the existing product or product range. Dynamic products occur when there are (or there are likely to be) conceptual design changes, the creation of new platforms for future incremental development, unstable market conditions with external technical advances and possibly regulatory change. Design strategies will place less emphasis on meeting the needs of existing production processes and there may be an associated willingness to subcontract manufacturing tasks. The categorisation of products into static and dynamic is therefore not precise, since it results from a weighing up of all the possible factors that contribute to product status (Hollins and Pugh, 1990, 34-35). This assessment was made by the research team on the basis of the empirical evidence collected.

Product Age   This is based on the first appearance of the product. Products going back to the pre-war period or earlier are categorised as *old*. Middle-aged or *middling* products are those at least 15 years old, but dating from a post-war period, while *young* products are less than 15 years old and typically less than 10 years.

Strategic Focus   This represents the particular strategic emphasis of the establishment in relation to the key value disciplines of *market*/customer intimacy; *operational* (manufacturing) excellence; and product/*technical* leadership (Treacy and Wiersema, 1993). These can be expected to influence the product development directions of the establishment and hence the product trajectory. The classification provides only a crude distinction between establishments, since for many companies all three foci are considered important to some degree. Nevertheless, previous research suggested that most respondents could indicate the relative importance of these strategic positions (Alderman *et al.*, 1996).

Leader/Follower   This indicates whether the establishment *leads* its industry/product niche in terms of product development or whether it tends to *follow* the competition. The distinction relates to Freeman's notions of offensive and defensive strategies in relation to technology (Freeman, 1982) and is closely related to issues of resource availability and the cost of developing new technology in the establishment's product area (Twiss, 1986). It is possible that claims to leadership may sometimes reflect a desire to lead rather than actual leadership.[6]

On the basis of these parameters, each product studied was classified as following either a static trajectory, a dynamic trajectory or an intermediate trajectory. Table 4.1 shows how this was done.[7] Static trajectories are typified by limited conceptual design input, slow rates of change and a focus on market or operational rather than technical competition. Dynamic trajectories are typified by continuing conceptual design development, rapid rates of change and a focus on technical leadership. Intermediate trajectories display elements of each type and thus fall somewhere in between.

Results

The results presented below are derived from the 38 case studies. These comprise varying levels of detail and depth of information. Brief illustrations of six of the best matched triplets are provided in Appendix 4.2. This section of the chapter provides a more general interpretation of these case study findings.

Analysis of Trajectories

The individual trajectory characteristics are considered in turn first of all.

Pace   Not surprisingly, given that the project was concerned with a mature and traditional industry, few of the products studied could be considered to be changing rapidly. Only two (both in the instruments sector) were experiencing significant changes every five years or less. Incremental change could be very rapid, however, particularly in engineer-to-order or customise-to-order markets (ETO/CMTO).

Table 4.1   Classification of trajectories

Score

Parameter	0	1	2
Pace	Slow	Medium	Fast
Design status	Static	Uncertain	Dynamic
Age	Old	Middling	Young
Technical leadership	Follower	Product dependent	Leader
Strategic emphasis	Operations	Market	Technical

Static trajectories score:	0-3	15 cases
Intermediate trajectories score:	4-6	15 cases
Dynamic trajectories score:	7-10	8 cases

Source: Alderman (1997).

Design Status   Table 4.2 shows the distribution of static and dynamic product designs. On the face of it there appear to be more dynamic products in the South East sample as anticipated and a higher proportion of static products in the North and West Midlands, but these differences are not significant.[8]

Table 4.2   Product design status

Design status

Region	Static	Dynamic
North	11	4
West Midlands	8	4
South East	4	7

Source: Alderman and Thwaites (1997).

Product Age   Relatively few of the products studied could be considered young (i.e. less than 15 years old). This is in part a consequence of looking at the engineering industry. Longevity of some products was a fact of life for many establishments and a problem for some in that it restricted the opportunities for strategic product portfolio management and rationalisation of parts counts, drawings etc. Customers often required that very old products be supported or wanted specials that harked back to very old designs. One or two establishments had a policy of not supporting old products beyond a certain stage, but for most this was not a sensible option given the ease with which customers could go elsewhere.

Strategic Focus   Table 4.3 shows the relative strategic emphasis adopted by the establishments in each region. It is clear that the hypothesised emphasis on technical focus in the South East compared to the other regions is not found. Indeed, a relatively high emphasis is placed on technical leadership in the Northern sample.

Table 4.3   Strategic focus

Focus (ranked first or equal first)*

Region	Markets	Technical	Operations
North	8	8	5
West Midlands	10	5	3
South East	6	4	3

* multiple response possible if ranks equal.
Source: Alderman and Thwaites (1997).

Product Development Leadership   About two fifths of the establishments studied regarded their approach to product development as one of leadership (i.e. attempting to be first in the market place with new or improved products). There were no significant differences on a matched pair basis between the regions. While many respondents considered that the particular product studied was a world leader in technical terms, this could not always be said to be true for the product portfolio as a whole. In other words, it tended to reflect the performance of the establishment in terms of one small niche within a much broader market base. For some establishments, small size and limited market power meant that a follower strategy was the only viable option for them. Contrary to the general hypothesis concerning the periphery, there were a number of establishments in the North that were clearly technical leaders in their particular field or market niche. However, this was not always reflected in market leadership, and one company in particular was highly inventive, but less successful when it came to innovation in the marketplace.

Table 4.4 shows how the matched establishments compared in terms of their class of product trajectories. On the face of it there appears to be some support for the basic hypothesis in that in the West Midlands there were no highly dynamic trajectories identified. Moreover, a visual comparison of the Northern and South Eastern matches suggests that the South East establishments either enjoyed similar or more dynamic trajectories than their Northern counterparts, again consistent with the hypothesis. However, these differences are not significant on a pairwise basis and the data in this form obscures the fact that some of the northern products actually had higher total scores than elsewhere.[9]

In aggregate, though, there are proportionately more static trajectories in the North than amongst the matched establishments in the other two regions put together (Table 4.5).[10] This does ignore the fact that some static Northern trajectories had higher scores than the matched static trajectories and that when the trajectory was dynamic in the North it was often considerably more so than in the other regions.

There is no strong evidence therefore, that, having matched establishments on key characteristics, there are any systematic differences in technological trajectory between the regions. There are examples of matched establishments that exhibit almost identical trajectories, while others differ substantially. In the former cases, the trajectories are being driven primarily by external factors, while in the latter internal factors would seem to be important.

Internal Factors

The internal drivers of the technological trajectory could be classified under three broad headings: an emphasis on cost reduction or design for manufacture; expansion or market-related (such as the deliberate targeting of new niches or increased market share); and technical performance criteria. Fully half of the establishments had in place cost reduction programmes as part of their

Table 4.4   Product development trajectories amongst the matched establishments

North	Trajectory	South East	Trajectory	West Midlands	Trajectory
N1	S	SE1	S	WM1	S
N2	S	SE2	S	WM2	I
N3	D	SE3	D	WM3	I
N4	S	SE4	I	WM4	I
N5	I	SE5	D	WM5	I
N6	S	SE6	I	WM6	I
N7	S			WM7	I
N8	S	SE8	S	WM8	S
N9	S			WM9	I
N10	S			WM10	I
N11	I	SE11	I		
N12	D			WM12	I
N13	S	SE13	I		
N14	D	SE14	D	WM14	S
N15	D	SE15	D		

S = static; I = intermediate; D = dynamic
Source: Original case study data.

Table 4.5   Static versus non-static trajectories

	Trajectory		
Region	Static	Non-static	Total
North	9	6	15
South East + West Midlands	6	17	23
Total	15	23	38

Source: Original case study data.

development activities. This study confirmed previous findings (Alderman *et al.*, 1996) concerning the importance of price as a competitive factor in the engineering industry. In contrast, only five establishments were clearly pushing product development from a performance or technical capability perspective. The results indicate that product development in Northern region establishments is just as likely, if not more so, to be driven by growth or expansion objectives as by cost competition. The corollary is that in the South East product development was just as likely to be driven by cost reduction pressures.

External Factors

External drivers depended heavily on the product market. Original equipment manufacturers (OEMs) were critical to many establishments, to the extent that in some product areas (automotive components being the classic example) product development activities were closely tied to the programmes of the OEMs. In other product areas legislative requirements, particularly CE marking and environmental directives, were important drivers.[11] There were some examples of products being withdrawn owing to the cost and difficulty of redeveloping them to comply with the new European legislation.

For companies serving CMTO or ETO markets, specific customer requirements were the principal external drivers on the development process. Where the OEMs were dominant in the market, there was a high level of regulation, or there was intense cost competition, the matched establishments tended to be responding to the same external pressures and the technological trajectories displayed very similar characteristics. These trajectories tended to be static, reflecting a risk-averse approach to innovation in the face of such dominant external forces. On the other hand, when the establishments addressed a highly specialised market niche, responded to ETO or CMTO market requirements, or were able to pursue technically oriented strategies, the trajectories tended to vary more, but in an idiosyncratic way.

External Networking

The networking thesis suggests that technological advance is a function of both internal and external resources. While internal resources varied considerably, there was no pattern to this on a pairwise (regional) basis. Neither were there systematic regional differences in the way that the matched establishments sought external technological inputs.

Overall, external networking was less prevalent than might have been expected. Only about one quarter of the establishments externally sourced some or all of the core technology in the product. In a number of cases this was through licensing-in or acquisition. A number of corporate establishments relied on corporate facilities for the core technology. Just under half of the establishments had key suppliers that provided some degree of design input, though not necessarily in relation to the core technology. There were no differences between regions on a pairwise basis in terms of external sourcing of technology, and product structure was a much stronger predictor of the need for external sourcing of technology.

The majority of the case studies demonstrated quite clearly a lack of local networking, or local systems of integration, in terms of the product development process. Where local networking occurred it was often fortuitous, in that the local University or a local supplier possessed, or had developed, the required technological capability, but there was no intrinsic networking relationship building up this mutually beneficial capability in the first place.

Another observation is that, while there were a few examples of the strategic use of the local innovation support infrastructure there were also examples of disillusionment with attempts to draw on local expertise and organisations or institutions in the local environment.[12] The most notable of these occurred in the South East, the region that, theoretically, is supposed to have the most conducive milieu. In general, local agencies were relied on for assistance with training provision, usually in relation to the production phase of specific developments, but not for product development itself. Most of the network relationships observed were established on a national or international, rather than local, basis and the inadequacies of the local supply base for key technological inputs to the development process were cited by establishments in all three regions.[13] For many of the South East establishments, preferential accessibility to national and international markets was cited as a positive location factor for companies, aiding the search for new technologies and the tapping into international customer or supplier networks.

The use of suppliers for technological inputs was not observed to the extent that the networking thesis might have predicted and the information collected indicates that most of these establishments were not particularly sophisticated in terms of systematically using their suppliers as a technological resource.[14] The case studies suggest that, in this set of establishments at least, it is the internal resources that constitute the primary source of development potential. Establishments in all three regions expressed concerns about how to capture and retain the tacit knowledge possessed by their engineering staff and the loss of key technical people was a constant worry.

In general, it was striking how insulated the establishments were from their local environment. At least as far as product development and the activities of the technical function were concerned, there was little influence of the local environment or business culture that could be discerned as having any kind of effect on the approach to product development or the resulting technological trajectories of the establishments' products. The case studies revealed that, rather than reflecting local conditions in the external environment, what is more important are the traditions within the company that lead to consistent responses to product development requirements. Product development procedures are frequently derived by capturing established practices. This was true even of externally acquired establishments, provided they had retained product development autonomy. In a lot of engineering establishments, therefore, one finds that product development follows a particular process, because that is how it has always been done. In a number of cases, changes to this process were seen to come about through the deliberate appointment of individuals who brought experience, not only from other companies, but from other (possibly more innovative) sectors.

Conclusions

This chapter has presented some results of a study that matched medium sized engineering establishments in an attempt to find evidence for systematic patterns of product development trajectory at the regional level. The principal conclusion from this work is that the evidence for such patterns is at best very weak. What comes across most strongly is the diversity of experience within any particular locality and the absence of any consistent regional patterns of trajectory, contrary to the initial hypothesis.

The case studies illustrated how the trajectory is influenced by both internal and external drivers, reflecting the strategic positioning of the establishment, its market and wider environmental context, its approach to external technology acquisition and the intrinsic characteristics of the technology itself. A striking finding is the degree of similarity between the technological trajectories of many of the matched engineering establishments in different regional contexts. In these situations external drivers, emanating from the product market or regulatory regime for instance, dominated the establishments' responses. However, these similar trajectories arose in the context of different sets of 'local' conditions.

The principal conclusion to emerge from this research then is that, in the engineering industry, local networks for technical development are not important; in fact in many instances they appear totally irrelevant. The concept of the innovative milieu appears, on the basis of these investigations, to have little bearing on the reality of how engineering establishments pursue product development. The critical issue from a networking perspective is the ability of the establishment to tap into both national and international networks of customers, suppliers and other third parties.

These findings are echoed by other recent studies undertaken in different industrial and locational contexts: for instance, Hardill *et al.* (1995) found no evidence of any use whatsoever of local institutional networks by firms in the textile industry in the East Midlands of the UK; Larsson and Malmberg (1997) concluded from their study of the Swedish machinery industry that there was no support for the idea that local (or even national) networking has a positive impact on the performance of firms; and Grotz and Braun (1997) similarly found limited evidence for local networking in the German mechanical engineering industry in relation to technical issues. Finally, the lack of a milieu effect in the South East of England, despite its possession of many of the crucial milieu attributes, is mirrored by research into technological award winners in Hertfordshire by Hart and Simmie (1997) which concluded that the innovative firms studied there did not engage in local networking and relied far more on national or international linkages.

The finding that for technological advance proximity does not seem to be important in the engineering context can partly be explained by the fact that, despite not being perceived as a 'high technology' industry, most of the products studied were highly advanced in technical terms and were dependent upon high technology components. The expertise required to design such components is not universal and frequently rare, so that the search for qualified suppliers or other sources of technical inputs is inevitably national or international in scope. Where such expertise is to be found locally this is fortuitous and may confer advantages of proximity, but it does not necessarily constitute a milieu effect.

The research shows that the presence or absence of local connections is insufficient to account for aggregate differences in regional innovation performance. For the engineering industry, such relationships with the local environment as were observed were found to be highly establishment specific and could not be predicted from general theoretical notions. Matched establishments exploited their local environment in very different ways. Local engagement was more often related to activities in support of technological development, such as training, rather than to technical advance *per se*. This is consistent with Tödtling's (1994) claim that the local environment and global networks perform different functions. This diversity of local engagement is a function of historical antecedents in the establishment and its environment, its strategic outlook, and the influence of external (global) drivers on the business.

The case studies have suggested that the culture and approach to product development within an engineering establishment results not so much from the local (business) environment as from the internal (corporate) culture of the enterprise, the embodiment of routines and procedures within the product development process and the accumulation of tacit knowledge and experience associated with a particular product and its applications. Technological trajectories display a high degree of path dependency precisely because the technical function becomes locked in to traditional ways of working that are difficult to break out of. It is questionable whether, in the type of establishment studied here, stimuli for change to product development approaches will arise from the local environment.

Acknowledgements

This research was funded by the UK Economic and Social Research Council under Grant No. R000234769. The research was conducted jointly with Alfred Thwaites of the Centre for Urban and Regional Development Studies (CURDS).

The extensive assistance of a large number of industrialists, managers and engineers is gratefully acknowledged, without whose willing participation this research would not have been possible. Thanks are also due to David Maffin who participated in some of the Northern region case studies and to Betty Robson for administrative assistance with the project. The comments of the participants in the Residential Conference of the IGU Commission on the Organisation of Industrial Space in Gothenburg on 'Technological Change in Space' are also gratefully acknowledged; however, the author alone is responsible for the analysis and interpretation in this chapter.

Notes

1 It was not feasible to match identically in terms of products (i.e. by identifying establishments that were direct competitors in each case), because many engineering products are tailored to specific niche markets and often there is only one UK manufacturer operating in that market niche. The product matching therefore took account of factors such as product complexity, since previous research had demonstrated the critical effects of such variables on the development process (Alderman *et al.*, 1996).

2 The population of establishments in the chosen industries is smaller in this region than the other two and therefore the opportunities for identifying replacements was correspondingly lower.

3 Matching problems included: an inability to identify any establishments with the matching criteria; refusal to participate; a failure to make contact with the right person within the organisation; cancellation of visits owing to crises within the business; or a lack of product development activity (see Appendix 4.1 for details).

4 The M25 is the London orbital motorway.

5 Design activity is not always internalised, as Walsh (1996) points out. The empirical work revealed that industrial design work concerned with form and aesthetic appearance (as opposed to engineering design, concerned with function) was often out-sourced. For dynamic products it is perhaps more likely that external sources of design expertise will be drawn upon, both in terms of engineering and industrial design.

6 It is worth noting that this indicator does not measure the same thing as the strategic focus. The strategic focus is more general, reflecting the strategic approach of the establishment and its relative emphasis on competing through technology, while the leader/follower measure is an assessment of how well the individual product stacks up against similar offerings from other manufacturers in terms of its technical characteristics.

7   The purpose of this exercise was not to create a robust quantitative score, but to provide a means of distinguishing broad categories of trajectory, taking into account the range of factors considered relevant. The scoring makes allowance for any uncertainty or ambiguity in the assignment of categories as reported in Table 4.1.

8   A pairwise test for differences between establishments in the North and those in the South East is only significant at the 10% level using the binomial approximation of the McNemar test.

9   The establishment with the most dynamic trajectory on these measures was in fact a Northern region one, while one of the two establishments with the lowest trajectory score was located in the South East. A Wilcoxon matched-pairs signed-ranks test is not significant at the 5% level for a two-tailed test.

10 A Sign Test (equivalent to the McNemar test) rejects the null hypothesis that the median of the differences between trajectories is zero at the 0.05 level.

11 CE marking indicates that a product complies with harmonized EU requirements for safety and health.

12 This was noted in one or two of the Northern region cases, partly because there is a well developed support infrastructure and a greater range of public policy instruments that the informed company can exploit. This was the strategy of case N12 with respect to its R&D activity, which it only instigated if external support was available.

13 This reflects in part a wider problem of the running down of the UK engineering base and the failure of UK companies to compete with foreign companies in many areas of technology.

14 This is consistent with analysis from a previous project which investigated, among other things, the degree of supplier involvement in product development projects (Alderman, 1996).

References

Alderman, N. (1986), *A Case Study of the Application of Log-linear and Logit Models in Industrial Geography*, unpublished Ph.D. thesis, University of Newcastle upon Tyne.

Alderman, N. (1994), *The Diffusion of New Technology in the British Mechanical Engineering Industry 1981-1993*, End of Award Report to ESRC (Grant no R000221136), Centre for Urban and Regional Development Studies, University of Newcastle upon Tyne.

Alderman, N. (1996), 'Networking for Product Development in Engineering', paper presented at the Sixth International Conference on Economics and Policy on Innovation: Networks of Firms and Information Networks, Cremona and Piacenza, Italy, 5-7 June 1996.

Alderman, N. (1997), 'Product Development Trajectories: Networks and Resources', paper presented at the VII Conference on the Economics and Policies of Innovation: Technical Progress and Innovation, Cremona, Italy, 11-13 June, 1997.

Alderman, N., Davies, S. and Thwaites, A.T. (1988), *Patterns of Innovation Diffusion: Technical Report*, Centre for Urban and Regional Development Studies, University of Newcastle upon Tyne.

Alderman, N., Maffin, D., Thwaites, A.T., Braiden, P., and Hills, W. (1996), *Engineering Design and Product Development and its Interface with Manufacturing*, End of Award Report to ESRC (Grant No R000290002), Centre for Urban and Regional Development Studies, University of Newcastle upon Tyne.

Alderman, N. and Thwaites, A.T. (1997), *Technological Trajectories at the Local Scale: Product Development in Engineering*, End of Award Report to ESRC (Grant No R000234769), Centre for Urban and Regional Development Studies, University of Newcastle upon Tyne.

Aydalot, P. (1988), 'Technological Trajectories and Regional Innovation in Europe', in P. Aydalot and D. Keeble (eds), *High Technology Industry and Innovative Environments*, Routledge, London, pp. 22-47.

Biemans, W.G. (1992), *Managing Innovation within Networks*, Routledge, London.

Braiden, P.M., Alderman, N. and Thwaites, A.T. (1993), 'Engineering Design and Product Development and Its Relationship to Manufacturing: A Programme of Case Study Research in British Companies', *International Journal of Production Economics*, vol. 30-31, pp. 265-272.

Chesbrough, H.W. (1994), 'Firm Level Technology Trajectories: Implications for Industry Innovation', paper submitted to the International Joseph Schumpeter Society 1994 Congress, Münster, Germany.

Cooke, P. and Morgan, K. (1993), 'The Network Paradigm: New Departures in Corporate and Regional Development', *Environment and Planning D: Society and Space*, vol. 11, pp. 543-564.

Dosi, G. (1982), 'Technological Paradigms and Technological Trajectories', *Research Policy*, vol. 11, pp. 147-162.

Dosi, G. (1988), 'The Nature of the Innovative Process', in G. Dosi, C. Freeman, R. Nelson, G. Silverberg and L. Soete (eds), *Technical Change and Economic Theory*, Pinter Publishers, London, pp. 221-238.

Freeman, C. (1982), *The Economics of Industrial Innovation* (2nd ed.), Pinter Publishers, London.

Gardiner, J.P. (1984), 'Design Trajectories for Airplanes and Automobiles during the Past Fifty Years', in C. Freeman (ed.) *Design, Innovation and Long Cycles*, Pinter Publishers, London, pp. 121-142.

Georghiou, L., Metcalfe, J.S., Gibbons, M., Ray, T. and Evans, J. (1986), *Post-Innovation Performance: Technological Development and Competition*, Macmillan, Basingstoke.

Grotz, R. and Braun, B. (1993), 'Networks, Milieux and Individual Firm Strategies: Empirical Evidence of an Innovative SME Environment', *Geografiska Annaler B*, vol. 75, pp. 149-162.

Grotz, R. and Braun, B. (1997), 'Territorial or Trans-territorial Networking: Spatial Aspects of Technology-oriented Co-operation within the German Mechanical Engineering Industry', *Regional Studies*, vol. 31, pp. 545-557.

Håkansson, H. (1987), *Industrial Technological Development: A Network Approach*, Croom Helm, London.

Hardill, I., Fletcher, D. and Montagné-Villette, S. (1995), 'Small Firms' "Distinctive Capabilities" and the Socio-Economic Milieu: Findings from Case Studies in Le Choletais (France) and the East Midlands (UK)', *Entrepreneurship and Regional Development*, vol. 7, pp. 167-186.

Harris, R.I.D. (1988), 'Technological Change and Regional Development in the UK: Evidence from the SPRU Database on Innovations', *Regional Studies*, vol. 22, pp. 361-374.

Hart, D. and Simmie, J. (1997), 'Innovation, Competition and the Structure of Local Production Networks – Initial Findings from the Hertfordshire Project', paper presented at the International Conference, 'Regional Frontiers', Frankfurt (Oder), Germany, 20-23 September 1997.

Hollins, W. and Pugh, S. (1990), *Successful Product Design*, Butterworth, London.

Johne, A. and Snelson, P. (1990), *Successful Product Development*, Blackwell, Oxford.

Larsson, S. and Malmberg, A. (1997), 'Innovations, Competitiveness and Local Embeddedness: A Study of the Swedish Machinery Industry', paper presented at the IGU Commission on the Organisation of Industrial Space 1997 Residential Conference on 'Technological Change in Space', Gothenburg, Sweden, 3-9 August 1997.

Lundvall, B.A. (1988), 'Innovation as an Interactive Process: From User-Producer Interaction to the National System of Innovation', in G. Dosi, C. Freeman, R.

Nelson, G. Silverberg and L. Soete (eds), *Technical Change and Economic Theory*, Pinter Publishers, London, pp. 349-369.

MacPherson, A.D. (1992), 'Innovation, External Technical Linkages and Small Firm Commercial Performance: An Empirical Analysis from Western New York', *Entrepreneurship and Regional Development*, vol. 4, pp. 165-183.

Maillat, D. (1990), 'SMEs, Innovation and Territorial Development', in R. Cappellin and P. Nijkamp (eds), *The Spatial Context of Technological Development*, Avebury, Aldershot, pp. 331-351.

Maillat, D. (1991), 'The Innovation Process and the Role of the Milieu', in E. Bergman, G. Maier, and F. Tödtling (eds), *Regions Reconsidered: Economic Networks, Innovation, and Local Development in Industrialised Countries*, Mansell, London, pp. 103-117.

Malecki, E. (1995), 'Culture as Mediator of Global and Local Forces', in B. van der Knaap and R. Le Heron (eds), *Human Resources and Industrial Spaces*, Wiley, Chichester, pp. 105-127.

Malerba, F. (1992), 'Learning by Firms and Incremental Technical Change', *Economic Journal*, vol. 102, pp. 845-859.

Massey, D. (1984), 'Introduction', in D. Massey and J. Allen (eds), *Geography Matters! A Reader*, Cambridge University Press, Cambridge, pp. 1-11.

Nelson, R.R. and Winter, S.G. (1982), *An Evolutionary Theory of Economic Change*, Belknap Press, Cambridge, Mass.

Northcott, J. and Rogers, P. (1984), *Microelectronics in British Industry: The Pattern of Change*, Policy Studies Institute, London.

Oakey, R.P., Rothwell, R. and Cooper, S. (1988), *Management of Innovation in High Technology Small Firms*, Pinter, London.

O'Farrell, P.N. and Hitchens, D.M.W.N. (1988), 'Inter-Firm Comparisons in Industrial Research: The Utility of a Matched Pairs Design', *Tijdschrift voor Economische en Sociale Geografie*, vol. 79, pp. 63-69.

Peck, F.W. (1985), 'The Use of Matched-Pairs Research Design in Industrial Surveys', *Environment and Planning A*, vol. 17, pp. 981-989.

Perrin, J-C. (1991), 'Technological Innovation and Territorial Development: An Approach in Terms of Networks and Milieux', in R. Camagni (ed.), *Innovation Networks: Spatial Perspectives*, Belhaven Press, London, pp. 35-54.

Phelps, N.A. (1995), 'Regional Variations in Rates and Sources of Innovation: Evidence from the Electronics Industry in South Wales and Hampshire-Berkshire', *Area*, vol. 27, pp. 347-357.

Piore, M.J. (1992), 'Technological Trajectories and the Classical Revival in Economics', in M. Storper and A. J. Scott (eds), *Pathways to Industrialisation and Regional Development*, Routledge, London, pp. 157-170.

Sharp, M. (1990), 'Technological Trajectories and Corporate Strategies in the Diffusion of Biotechnology', in E. Deiaco, E. Hornell and G. Vickery (eds),

Technology and Investment: Crucial Issues, Pinter Publishers, London, pp. 93-114.

Sternberg, R. (1997), 'Innovative Linkages and the Region – Theoretical Assumptions versus Empirical Evidence', paper presented at the IGU Commission on the Organisation of Industrial Space, Residential Conference on 'Technology in Space', Gothenburg, Sweden, 3-9 August 1997.

Takeuchi, H. and Nonaka, I. (1986), 'The New New Product Development Game', *Harvard Business Review*, vol. 64, 1, pp. 137-46.

Thwaites, A.T., Oakey, R.P. and Nash, P. (1981), *Industrial Innovation and Regional Development*, Final report to the Department of the Environment, Centre for Urban and Regional Development Studies, University of Newcastle upon Tyne.

Thwaites, A.T., Edwards, A. and Gibbs, D.C. (1982), *Interregional Diffusion of Production Innovations*, Final Report to the Department of Trade and Industry and Commission of the European Communities, Centre for Urban and Regional Development Studies, University of Newcastle upon Tyne.

Tödtling, F. (1990), 'Regional Differences and Determinants of Entrepreneurial Innovation – Empirical Results of an Austrian Case Study', in E. Ciciotti, N. Alderman and A.T. Thwaites (eds), *Technological Change in a Spatial Context*, Springer-Verlag, Berlin, pp. 259-284.

Tödtling, F. (1994), 'The Uneven Landscape of Innovation Poles: Local Embeddedness and Global Networks', in A. Amin and N. Thrift (eds), *Globalisation, Institutions, and Regional Development in Europe*, Oxford University Press, Oxford, pp. 68-90.

Treacy, M. and Wiersema, F. (1993), 'Customer Intimacy and Other Value Disciplines', *Harvard Business Review*, vol. 71, 1, pp. 84-93.

Twiss, B. (1986), *Managing Technological Innovation* (3rd ed.) Pitman, London.

von Hippel, E. (1988), *The Sources of Innovation*, Oxford University Press, New York.

Walsh, V. (1996), 'Design, Innovation and the Boundaries of the Firm', *Research Policy*, vol. 25, pp. 509-529.

Wheelwright, S. and Clark, K. (1992), *Revolutionizing Product Development*, Free Press, New York.

Appendix 4.1   Case-wise Matching of Establishments

Northern region	South East	West Midlands
N1	SE1	WM1
N2	SE2	WM2
N3	cancellation SE3	refusal refusal WM3
N4	SE4	non-response no product development WM4
N5	SE5	refusal WM5
N6	SE6	WM6
N7	no match	WM7
N8	refusal SE8	refusal WM8
N9	no match	WM9
N10	non-response	WM10
N11	refusal SE11	refusal refusal
N12	non-response	non-response WM12
N13	SE13	no match
N14	SE14	WM14
N15	SE15	no product development cancellation
N16	cancellation	non-response non-response
N17	no match	no match
N18	no match	no match
N19	no match	no match
N20	no match	cancellation

Successful case studies are labelled N1, SE1, WM1 ... etc.
No match = no establishment with required characteristics identified.
Non-response refers to establishments identified but where contact with the
appropriate respondent failed after up to a dozen attempts.

Appendix 4.2   Illustrative Case Study Summaries

The following brief summaries for six of the matched triplets illustrate the diversity of experience observed during the case studies as well as some of the important factors influencing the nature of technological trajectories. Further details are provided in Alderman and Thwaites (1997, Appendix 1). The case study identifiers correspond to those in Appendix 4.1 and Table 4.4 above.

Triplet 1 (N1, SE1, WM1)

This triplet of medium to large establishments exemplified the similarities in approach to product development in the face of conservative customers with strong safety/reliability requirements within a highly competitive global market and a national labour market for technical/engineering skills. Different strategic foci, corporate environments and technical resources gave rise to similar product trajectories. High cost penalties arising from product failure, either during test or in the field, promoted a risk aversive approach to development and strictly incremental change based on tried and tested technologies.

The triplet was also characterised by a reliance on accumulated and tacit knowledge. Transfer of personnel within the industry (and between regions) was seen to be an important component of technological transfer, and awareness of industry developments was enabled through common membership of standards committees, Trade Associations and the like. All three establishments exhibited a high degree of vertical integration and only limited reliance on key external suppliers.

Triplet 2 (N2, SE2, WM2)

This triplet of small corporate establishments illustrated the dominant role of original equipment manufacturers (OEMs) in certain subsectors of engineering. All three operated development programmes that were geared to the timetables and needs of the OEMs. It was difficult for these establishments to be radically innovative, since they were dependent on the OEMs for their market. Differences in trajectory within this triplet were attributable primarily to differences in corporate structure. All three establishments operated in such a way that they appeared to be largely insulated from their local environment. With the exception of WM2, these establishments were not located close to the major OEMs; however, in the latter's case the location decision was primarily to do with distribution considerations (i.e. position relative to all customers or potential customers) rather than proximity for product development collaboration.

Triplet 3 (N3, SE3, WM3)

This triplet of small instrumentation establishments with dynamic products illustrated how different approaches to the organisation of the product development process were successful in different contexts. All three required considerable external technological inputs, but approached technology acquisition in different ways: acquisition of competitors with potentially superior technology (N3); collaboration with a key supplier (SE3), located outside the UK, in order to obtain microelectronics expertise perceived not to exist in the UK; and use of local Universities for technical support (WM3). Each establishment had some kind of engagement with its locality, but this also differed markedly. N3 relied on the local environment primarily for training provision. It preferred to build up in-house technical capability, particularly as it moved into electronics and software. SE3 had considerable interaction with customers, suppliers and even competitors, this reflecting the greater density of such organisations within that part of the South East compared to the North or West Midlands. WM3 drew on local technical expertise, primarily through Higher Education Institutes (HEIs), but also participated in national and European technical programmes. Its key suppliers were non-local, however.

The organisation of product development differed markedly between the three sites. N3 had moved to highly structured product development procedures with cross-functional project teams as a result of engaging an international firm of consultants to advise it. SE3 operated a traditional functional structure with product development retained within the engineering function. Recent changes in product development at SE3 had been primarily a result of the recruitment of a new Technical Manager who had instigated changes to what was a well established and ingrained internal culture of product development. He brought with him experience and practices from a different industry sector. WM3 utilised a matrix form of organisation with the active involvement of the Managing Director, who essentially drove the product development process and personally headed up the small R&D department.

Triplet 4 (N4, SE4, WM4)

This triplet of small, autonomous or independent establishments manufacturing machinery reflected different ways of serving the marketplace, ranging from a reliance on a standard product (N4) to a focus on specials (WM4). Differences in technological trajectory reflected these market relationships and the strategic foci of the establishments, with different internal drivers of the product development process: value engineering and cost reduction versus expansion of the product range, the latter giving rise to a more dynamic trajectory (SE4).

Restrictions on the innovativeness of N4 were seen to lie with specific skills areas such as microelectronics expertise. N4 had tried to overcome this through a Teaching Company Scheme with a local HEI and it relied on suppliers of control systems for most of the technology in this area. This was also the principal external technology for SE4. No key suppliers existed within the local area and the company perceived a more general problem to exist in terms of the UK supply base, such that suppliers often had to be sought abroad. The company's philosophy on product development was very much an in-house one. Attempts to use external agencies, such as a local college, to inform the development process were perceived to have had limited success, largely due to the lack of external understanding of the company's markets.

WM4 subcontracted manufacturing activity locally; key suppliers (primarily for the control systems) are not found locally and for some specialist components there were only one or two UK manufacturers. Since WM4 operated in an industry that has seen considerable decline and job shedding, there were a lot of individuals on the labour market with design expertise and it sometimes subcontracted design activity to suitably qualified individuals operating as self-employed design consultants. This is something that the local labour market of WM4 permitted rather more readily than that of either N4 or SE4 which, historically, have not had concentrations of this type of engineering activity.

Triplet 5 (N5, SE5, WM5)

This triplet of medium to large corporate establishments illustrated the importance of the supply base, particularly first tier suppliers, in providing much of the technology embodied in a comparatively complex product. All three establishments operated fairly sophisticated supplier selection and approval procedures. With a couple of exceptions, key suppliers were not local and, where they were, this was largely a serendipitous outcome. Local suppliers tended to be more important for the subcontract of manufacturing activity, but not in relation to the product development process, where national, and increasingly international, technical supply networks were important.

N5 retained core competences in the critical performance and quality-related technologies. Most other technical requirements were out-sourced, such as rotating machinery and valves. The bulk of these were bought in as standard items. It had a policy of looking to local providers (particularly for subcontract manufacturing). Some disappointment was expressed concerning the response of local industry and N5 had found that, owing to the detailed nature of the specifications it places upon its suppliers, many potential suppliers were reluctant to handle contracts for it. For many of the more specialist supply items N5 had to look outside the region.

WM5 did not develop many of the core technologies itself and relied on its suppliers for the principal components. Most standard components were sourced from outside the region (e.g. from the South East or mainland Europe – despite a policy of UK sourcing), with the exception of the equipment's controller, which was designed to technical specification by a local supplier. This proximate relationship appeared to have been largely serendipitous, but owing to the need for close supplie involvement during the development phase, this local relationship was highly beneficial.

Triplet 6 (N6, SE6, WM6)

This triplet of small engineering establishments with a simple traditional product illustrated how intrinsic limits to the technology resulted in slow trajectories of development of a highly evolutionary nature. N6 was most constrained in this way and this was reflected in the fact that both SE6 and WM6 engaged in more conceptual development work and relied less on pre-existing designs to develop the product. N6 was also more internally oriented. Both SE6 and WM6 were willing to use external design houses when the internal resource became inadequate for the job in hand. N6 expressed the view that, in terms of technical support, its peripheral location was something of a handicap.

5 Innovation and Local Development: The Neglected Role of Large Firms

Jerry Patchell, Roger Hayter and Kevin Rees

Introducing Large Firms

In the US they have been called 'little giants'; in Germany, 'hidden champions'; and in Japan, 'backbone firms'. Firms in the range between small and medium- sized enterprises (SMEs) and giant multinational or transnational corporations (MNCs or TNCs) – firms which for simplicity's sake we call large firms (LFs) – exert a tremendous influence on their economies. They are at the cutting edge of introducing product, organisational and technological innovations; rearranging the local space economy; and re-invigorating regional, national, and global industrial organisation. Perhaps more than any other firm classification, LFs illustrate the importance of making connections. They combine ideas from a diversity of sources to create product, process, and organisational innovations that invigorate not only the LF but also the firms they trade and compete with.

LFs do not innovate in isolation. Hayter (1997) has begun to theorise the connectivity between factory, firm, and production system (FFPS); in this paper we go one step further by showing the connectivity of a firm, its factories and production system within its industry. A firm does not merely occupy a segment of an industry, as it engages competitors as well as suppliers and distributors. LFs perform a critical role in this connective engagement because they are the ones most actively developing a strategic competitive advantage: it is the only way they can break out of the small-firm sector and compete with MNCs. We will show the impacts that these LF strategies can have on the geography of production and their industries.

In the next (second) section, the paper introduces LFs and the role they play in advanced economies. The discovery of LFs causes a reconsideration of the theory of business segmentation used in geography. We argue that LFs constitute an important segment of firms which are conceptual *terra incognita* within industrial geography. However, rather than simply drawing attention to an ignored segment of the space-economy, we illustrate the role of LFs in breaking down the barriers between segments and invigorating an industry. We begin this process by highlighting Nakamura's (1990) depiction of the role of LFs in the Japanese economy. He along with Kuhn (1982, 1985) for the US and Simon (1992) for Germany are largely responsible for discovering the role of LFs. In their eyes, LFs are dynamic innovators that share some but not all the characteristics of SMEs and MNCs. Furthermore, they emphasise that LFs are an important feature of technically advanced, diversified industrial economies. In Canada, in comparison, Steed (1982) argues that LFs, which he calls 'threshold firms', are under-represented and need to be promoted for Canada to gain technologically-based competitive strengths. We argue that geography has to create a new theoretical construct to deal with their influence.

In the third and fourth sections of our paper we clarify the strategic and spatial characteristics of LFs, and finally in the fifth section we examine the ramifications of LFs upon the health of an industry, specifically the Japanese robot industry. The role of LFs as innovators is vital to recognising these characteristics, particularly in how innovation prompts LFs to engage with their local and global environments in the search for new sources of advantage. Those connections in turn induce further innovation in developing sophisticated production systems and networks of factories. Thus, LFs are local, to the extent that they remain attached and embedded to home places, and LFs are global, to the extent that their growth has internationalised activities. Moreover, LFs are big enough to exert noticeable local impacts. They are literally 'big firms locally' and as such have potentially special roles in linking the fortunes of localities with the forces of globalisation.

From Static Segmentation to Dynamic Industrial Organisation: The Role of Large Firms

In crude empirical terms, as a category in-between SMEs and MNCs, LFs occupy a substantial size range, roughly between 500 and 5,000 employees and between $50-100 million to $2-$3 billion in sales. However, neither this categorisation, nor estimating their numbers in the economy or people employed (there are few appropriate statistics) are useful in defining the dynamic role of LFs. We will rely on Nakamura's description of them in the Japanese economy to illustrate that role.

Nakamura's analysis identifies three stages in the evolution of LFs since the 1950s. First, in the 1950s and 1960s, LFs provided a key role in introducing new organisations and labour relations strategies to facilitate the introduction of new technologies. In a variety of industries, a few pioneering LFs, including Honda and Sony, revolutionised existing small firm (SF) management thinking by: focusing on products with growth potential; achieving the benefits of mass production and quality improvements in specific market niches; rapidly introducing new process technologies; and introducing meritocracy and egalitarianism in the workplace. According to Nakamura 1990, p. 8), the leading LFs of the 1960s distinguished themselves from SFs by achieving scale economies and high wages while providing critical support for the advances made by large firms in the assembly and new materials industries. There were different types of LFs. Thus machine tool LFs focused on developing internationally competitive products; regional LFs in various industries (for example, laundry, raincoat, miso, ham and sausage, furniture) developed national identities; and LF subcontractors in auto and other assembly industries emerged by improving the precision and standardisation of their products as well as improving delivery dependability.

Second, Nakamura (1990, pp. 12-17) argues that in the 1970s LFs led diversification and economic restructuring processes in the economy as a whole. In this period, leading Japanese LFs developed new applications and products in related or entirely new directions either through original R&D or by introducing technology developed elsewhere. One key strategy, often based on R&D-based scope economies, involved developing high-variety, small-lot production to supplement mass-production/distribution systems serving saturated markets (Nakamura, 1990, p. 17). Experience in developing high-variety/small-lot production also prepared the way for the factory automation systems of the 1980s.

Third, Nakamura (1990, pp. 19-29) argues that in the 1980s and 1990s LFs led the shift in the Japanese economy from its heavy industry and mass production bias towards integration in a more globalised division of labour in which Japanese firms elaborate niche markets, increasingly bridge manufacturing with primary and service activities, and create entirely new businesses. Within the general context of the information and communication techno-economic paradigm, there has been tremendous growth of highly differentiated markets as a result of the breakdown of the mainframe computer which created demands for a wide variety of peripheral devices and software, the shift towards flexible mass production which created demands for a wide variety of automated products and machines, while information technology also opened up new markets in such fields as records, ticketing and music/video rental. In addition, Nakamura argues that LFs in Japan are leading the way in terms of offering Japanese consumers new life style choices in clothing, record rentals, computer communications and corporate vision.

According to Nakamura (1990, pp. 27-29), leading-edge LFs in the 1980s differed from those in the 1960s . While LFs throughout the period depend on entrepreneurial ability to create businesses, the entrepreneurs of the 1960s were more likely to be 'lone wolves' whereas as the entrepreneurs of the 1980s were more willing to collaborate, merge and network with other firms of similar size and power. Japanese LFs have typically been closely involved in subcontracting relationships (and in a social division of labour), far more so than their counterparts elsewhere. In the 1980s, however, LFs engaged in alliances with other firms that proactively define strategic directions. In this regard, Japanese LFs have become more interested in diversifying their activities and markets by bringing together related technologies within strategic alliances or by internally using existing technological expertise to promote new products. In addition, the LFs of the 1980s allowed employees more input in the direction and organisation of the firm, so that the firm becomes more than the entrepreneur's vision. Indeed, for Nakamura, the LFs that become the giants of the 1990s will draw their competitive advantage less from economies of scale, and more from economies of scope and networking with other firms to enhance their innovativeness.

Kuhn and Simon present similar discoveries of the role of LFs in the American and German economies. However, these discoveries have made no impact in geography despite the increasing recognition of the importance of small firm networks, flexible specialisation, outsourcing, and the demise of Fordism, that have taken on increasing importance since the publication of Piore and Sabel's (1984) *The Second Industrial Divide*. We believe that the failure to recognise the role of LFs results from a commitment to a dual business segmentation model that depicts a void between SMEs and MNCs.

The dual business segmentation model rejects the idea that all firms compete for similar market opportunities (Averitt, 1968; Berle and Means, 1932; Galbraith, 1951, 1967; Robertson, 1928; Veblen, 1932). To briefly re-capitulate its essence, in the dual model, firms are divided into two, essentially non-competing segments, comprising *planning system* and *market system* firms, respectively. In contrast to the market model, in which independent

SMEs ensure the operation of the economy in accordance with broader societal wishes (the sovereignty of consumers), in the segmented economy MNCs are the dominant organisations that shape consumer tastes, supplant arm's-length relationships with internal, hierarchical transactions, and ultimately control the structure and rewards of the economy as a whole (Taylor and Thrift, 1982a, 1982b, 1983). In this view, the differences between planning and market system firms, in terms of size and scope, behavior and the markets in which they operate, are preserved by barriers to inter-segment mobility, particularly the implausibility of SMEs becoming MNCs. These barriers, which take concrete form as global production and marketing networks, financing, and the scale and scope of R&D, directly reflect the power of MNCs. Any competition offered by SMEs, specifically by loyal opposition firms, is highly localised and readily countered by MNCs. Indeed, the main relationships between market system and planning systems firms are ones of dependency, as the latter exploit the former for low- cost goods, outlets for markets and as acquisition targets, as needed. That is, planning system firms control market system firms.

In reality, patterns of segmentation are more complicated than those presented in the original dual business segmentation model. Even in considering relations between MNCs and SMEs, the dual model over-simplifies the diversity of ways to organise factory, firm, and production system. The absence of LFs within contemporary models of business segmentation and location, in our view, requires re-examining the diversity and changing nature of segmentation processes. While the size distribution of firms implies that LFs are an important empirical reality, their structure and performance has been virtually ignored in the geography of firms and in related regional development and production systems literature. In this sense, LFs are *terra incognita*. More importantly, the connections they make are unrecognised. Yet, it may well be that LFs are becoming more important. Moreover, in theoretical terms, as a transitional segment between the polar cases of SMEs and MNCs, the contemplation of LFs raises questions about the processes as well as the patterns of business segmentation. While the dual model is presented as a static pattern of segmentation, LFs have to be understood as dynamic (innovative) organisations. Indeed, a significant issue underlying discussions of LFs is the apparently conflicting impulses of segmentation (and the erection of barriers) and innovation (and the breaking down of barriers). Ultimately, any revised model of the segmented economy must address this conflict.

In the meantime, it should be recognised that the implication of LFs is to simultaneously define a business segment and highlight the permeability of segment boundaries. Thus, LFs blur processes of segmentation by suggesting a connection between SMEs and MNCs, beyond the 'long shadow' cast by MNCs over SMEs. LFs exist because of abilities to overcome the substantial market and internal organisational barriers that limit the growth of SMEs, while also resisting hostile attention from existing MNCs. Moreover, every LF is a legitimate threats to MNCs, threatening to become another MNC rival. This argument does not imply that mobility between segments is invariably linear from SME>LF>MNC (or, in more disaggregated fashion, SF>MSF>LF>MNC>TNC). Firms can become smaller (and different) as well as bigger (and different) and it may well be, for example, that competitive pressure from LFs are stimulating MNCs to change their behavior. More precisely, LFs often evolve from SMEs but they can originate as spin-offs from MNCs. Over time, LFs may become MNCs, be acquired by existing MNCs, downsize to SMEs, fail, or remain as LFs. However, LFs are not generated from a random process. As Kuhn (1985), Nakamura (1990) and Simon (1992) emphasise, LFs exhibit organisational, technological and product-market characteristics which, in combination, distinguish them from SMEs and MNCs.

LFs also encourage a re-appraisal of the links between segments. Indeed, the presence of LFs indicates that the original segmentation model over-emphasised barriers to mobility and the dominance of the 'power networks' of MNCs over SMEs. If MNCs do impose powerful barriers to the growth of smaller firms, the emergence of LFs must be based on some competitive advantage not readily (or already) dominated by MNCs. The most obvious source of such competitive advantage is know-how-based innovations. MNCs can effectively employ patents and brand names to protect their interests, but only for existing know-how. MNCs can acquire the know-how of smaller firms, including LFs, but only if the latter wish to sell. Alternatively, MNCs may try to duplicate the know-how generated by LFs, but such efforts may be costly, uncertain and ultimately less efficient. More broadly, public-sector education and training and the rapid growth of the information society are (massive) forces countering the monopolisation of knowledge for private purposes by MNCs. LFs can learn and innovate in ways that are not dictated solely by MNCs.

Within the context of inter-segment relations, therefore, the power networks that are largely controlled by MNCs are paralleled by learning processes that allow LFs the know-how to act independently of MNCs – as leaders in markets which MNCs have not entered or are too small for them to bother with, as competitors against MNCs in niche markets, and as component suppliers to MNCs. Indeed, learning (and innovation) raises all kinds of possibilities for mutual gains among firms which are based on cooperative as well as competitive relationships (Patchell, 1993a, 1993b). As recent analyses are beginning to comprehend, the manner in which firms compete, cooperate and control themselves takes on different permutations in different parts of the world (Glasmeier, 1988; Patchell, 1993a, 1993b; Storper, 1993). It is nevertheless difficult to generalise about the relative importance of power networks vis-à-vis learning networks in the context of business segmentation. To an important degree, the privileging of power networks, for example in the original segmentation model (Taylor and Thrift, 1983) and related literature (Amin and Robins, 1991), or the privileging of learning networks, for example in the flexible specialisation literature (Cooke and Morgan, 1993) is a matter of ideology, judgment and persuasion. There is no obvious mechanical rule to prove the dominance of one network over the other. Indeed, if we can be permitted some license, if the power of MNCs is said to be evident by their control over half of the world's economic assets, then the potential of SMEs (and LFs) is surely reflected in their control of the other half! If the metaphor of the shadow is relied upon to extend the power of MNCs over SMEs (and LFs) then we should remember how shadows vary over time and space, and are sometimes not seen at all.

An alternative view is to conceptualise power networks and learning in networks as interdependent (Hayter and Edgington, 1997). If firms bargain to gain access to information and prior know-how, then learning and know-how facilitate bargaining. Similarly, firms can learn internally or with others (Table 5.1). But these attempts also imply various kinds of bargains. In an intra-firm context, for example, the progressive enhancement of organisational know-how on the basis of learning by doing, the development of enterprise-specific skills which relate to the width and depth of experience of workers (Patchell and Hayter, 1995) and in-house R&D require negotiation of more complex work arrangements and coordination and systems of rewards and penalties. Similarly, the progression of inter-firm learning through learning by using, the relation specific skill and strategic R&D alliances invoke bargains with other firms (consumers, suppliers, rivals). In the case of the relation-specific skill, which defines how suppliers serve the needs of core firms within industries such as autos, electronics and robots, learning and the creation of value is closely tied to bargains between core firms and suppliers regarding price, quantity and general responsibilities (Asanuma, 1989; Patchell, 1993a, 1993b).

Table 5.1   Types of intra-firm and inter-firm learning

Intra-firm learning		Inter-firm learning	
Type	Basis	Type	Basis
Learning by doing	Job specialisation and machine-specific skills	Learning by using	Collaboration with consumers
Enterprise-specific skills	Width and depth of job experience	Relation-specific skills	Collaboration with suppliers
In-house R&D	Product and process innovation (minor and major)	Strategic R&D	Collaboration between in-house R&D groups for 'frontier' innovation

Learning within and among firms, and the implied bargaining, are not mutually exclusive choices. In most situations one form of learning will stimulate the other to some degree. However, firms can invest considerably in intra-firm learning without implying the same commitment to inter-firm learning. Indeed, problems of trust, the need for secrecy and the related costs and uncertainties of negotiating external contracts encourage just such a strategy. On the other hand, it is less plausible to think of inter-firm learning in the absence of significant intra-firm learning. Such arrangements have to be based on mutual contributions of know-how as well as trust. Moreover, if inter-firm learning is more difficult to organise, such externalised relationships generate stronger local spin-offs in the form of quality as well as quantity multiplier effects, in addition to greater long-term adaptability or flexibility. Around the globe, if forms of inter-firm learning vary in intensity and type, Japan is undoubtedly the leading exemplar and benchmark (Fruin, 1992).

In the segmented economy, interaction of power and learning networks necessarily implies different kinds of inter-segment relations than originally conceived (Figure 5.1). Rather, a firm-based strategy is the stimulus to connect with both local and global influences that are the firm's sources of learning and the bases on which to create the innovations that will become its competitive advantage. Porter (1990) has argued that the ambitious firm will seek

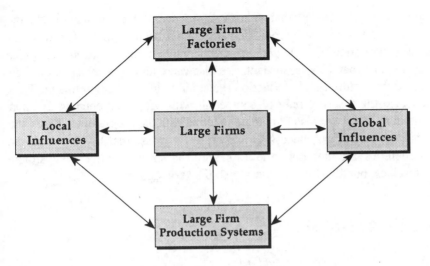

Figure 5.1   Strategic connections of large firms

advantages from the local culture, competition, institutions, markets, and suppliers from amongst which it arises. In his discussion of ubiquitification, Maskell (chapter 2, *this volume*) similarly emphasises the importance of such factors as sources of local competitive advantage. As our examples show, LFs are essentially products of these various influences in their local environments, but it is also obvious that they have achieved superior status because they have combined local influences with the discipline of global competition and standards, absorbed different cultures, buy from the best suppliers anywhere and, foremost, are reaching out to international markets.

LFs have no monopoly on learning from local and global influences. LFs, however, are both powerful and smart enough to break through SME barriers to growth and to challenge MNCs. LFs erode the static lumpiness of the traditional segmentation model because they define their own factory networks and production systems, apply effective competition to MNCs, and supply all competitors with essential components. What holds these firms together as a category is not their size *per se*, but common niche-opening strategies that take them from SME to giant-tamer.

The strategic connectivity of LFs impacts directly on the geographical distribution of firm activity. The home base of the firm supplies it with many of the influences propelling its advance. At the very least, the LF may be embedded in its home place as the locale where its entrepreneur has established and combined the factors that make the firm successful, but it may also be the region where an intricate production system has been nurtured. On the other hand, exploitation of scale and scope economies in the global market-place may require that LFs locate factories elsewhere and seek to replicate their home-base factories and production system in a different environment. Thus LFs face the task not only of competing with MNCs or opening up their internal markets, but also of coming to terms with the geographical bases and ramifications of their success. However, to the extent that LFs succeed, the beneficiaries are not only the LFs and the locales where they situate themselves, but also the industry which they reinvigorate.

Strategic Connections

Positioning in Industrial Structure

The common features shared by LFs are competitive strategies which define their long-run 'positioning' within particular industries and business sub-segments or types. Traditionally, strategies have been widely referred to in terms of forms of expansion (horizontal and vertical integration, internal and external growth). However, Porter's (1990, p. 34) analysis of strategy formulation best describes the activities of LFs. According to Porter, industry structure (defined as the underlying economic and technological characteristics of an industry, which determine the extent to which it allows enhanced profitability) and positioning within that structure (to find the most advantageous niche for a firm) are the two central concerns underlying competitive advantage. Although different industries may be more profitable than others, LFs in particular can find advantageous positions in any industry through a precisely configured strategy. Porter (1985) claims that cost leadership, differentiation (of products), a focus on reducing the cost of a limited range of products, and a focus on differentiating a small range of products comprise the range of strategic choices open to any firm. Porter does not venture to link these strategies with firm size or maturity. But in the general terms of corporate strategy, the firms that attempt to offer a full range of higher-value differentiated goods or higher- volume low-cost goods are MNC giants like General Motors, Philips, or Heinz. Some of these MNCs even try to supply a range of products that cover high, middle, and lower ends of the market. On the other hand, LFs break into these markets with strategies that focus primarily on differentiation in niche markets and secondarily on cost leadership in a limited number of goods. The efficiency and dynamism of LFs are rooted in *a continuing commitment to innovation*. The most successful LFs do not rely on a single innovation but develop several sources of innovation – from building relations with suppliers to enhanced manufacturing capabilities to distribution and marketing. From this perspective, the metamorphosis of LFs as innovators implies several organisational and product-market innovations. In so doing, LFs mediate forces of globalisation and localisation by organising a division of labour among value links that are organised internally and externally.

In terms of industry structure, LF strategies are shaped by the industrial and business sub-segments in which they operate. Markets for some consumer goods are both global and high-volume, whereas others, while still global, allow only limited production, and still others are only regional and characterised by a range between high and low volumes. Capital goods markets also offer this range of markets, but ratcheted down in volume and requiring different marketing skills. Primary and intermediate materials, parts, components, and services fit into these markets in the appropriate manner, with the same volumes and limitations of markets, but a major difference is that suppliers can offer their products to any number of end-use firms.

Within a very wide range of technological and market opportunities and constraints, innovation is a general characteristic of LFs. LFs are inherently leaders and are far more powerful sources of competition to MNCs than SMEs. Within end-use markets LFs have the choice of competing with MNCs head-to-head (based on some technological or organisational advantage) or, more likely, to create a niche that the MNCs have not realised. Markets of limited volume inherently are fertile grounds for the specialised efforts of LFs. To succeed in these markets LFs must match innovation with the appropriate blend of scale and scope economies and sophisticated management. That is, scale merits are a vital component of LF competitive strategy. Indeed, in practical terms, it is doubtful whether an SME can mount effective opposition to MNCs – without becoming an LF. However, out-competing giant end-use product firms in niche markets is not the only basis of LF market penetration. Perhaps more important to the strengthening of an industry is the LF role as component suppliers. End-use firms concentrate their efforts on product design and assembly and can realistically concern themselves with intensive development of only a few key components. That leaves substantial leeway for LFs (or SMEs that could become LFs) to improve the performance of any component, to combine parts into components for ease of manufacturability, or to reduce costs. Furthermore, whereas end-use firms are reluctant to sell superior components to competing firms, component supplier LFs have no such conflicts and can use the market supplied by several purchasing firms to achieve the scale and scope economies not available to any end-use firm making components in-house. In the past there has been a tendency to describe these component suppliers as dependent on end-use firms. However, there are different type of subcontractors (Holmes, 1986) and the emphasis of LFs as innovative differentiators is consistent with specialist subcontracting. They clearly are not in the same position of dependency as capacity subcontractors.

Existing LFs are not necessarily highly competitive. After all, innovative environments are uncertain, innovative capabilities cannot be guaranteed, and LFs cannot rely on the same financial resources or lobbying power as MNCs to bail themselves out. It may be speculated, however, that under contemporary globalisation forces, LFs are unlikely to remain laggards for long before failing or becoming SMEs. Perhaps in the less competitive times of the 1950s and 1960s laggard LFs could survive longer. Possibly, such laggards constituted a structural weakness in regional economies that suffered from de-industrialisation in the 1970s and 1980s?

Supported by economies of scale and scope in relation to closely integrated R&D and marketing efforts, LFs target specialised product markets, including high income market niches, on an international scale. The role of LFs as global niche producers is underlined by Kuhn (1982; 1986), Nakamura (1990) and Simon (1992), respectively for the US, Japan and Germany. As Nakamura (1990, p. 3) and Simon further note, LFs have achieved dominant national and even global positions in highly differentiated markets, a characteristic which distinguishes them from most SMEs. Simon (1992), for example, cites 25 German-based LFs, out of 39 interviewed, that were either the world's leading or second-leading producer of particular products, circa 1990. One firm (Gerriets) was the only producer in the world of its specialty while several firms enjoyed global market shares several times larger than that of their nearest rival. Korber/Hauni, for example, had 90 percent of the world market for cigarette machines throughout the 1980s. Overall, the 39 firms interviewed had average sales of $303 million, 2,904 employees, a world market share of 22.6 percent and a European market share of 31.7 percent (Simon, 1992, p. 116). The highly specialised, global market focus of LFs is further illustrated by the 50 'little giants' listed by *Business Week* (Hayter, 1997, pp. 247-248). In 1992, although specialised, the sales of these 50 firms were in the $200 million to $1.3 billion range and only five received less than

25 percent of sales from foreign markets (and only one less than 20 percent) and all but eight had factories in foreign countries.

The various roles of LFs as component suppliers, as competitors to giant firms or even as leading firms in smaller product or regional markets are not cast in stone. There is constant reassessment of strengths and reformation of strategy. Superior production of a key component may induce a component supplier to enter competition against end-use firms or refrain or devolve from end-use production. To cite some well-known examples, Honda's performance in engine manufacture gave it the door to enter automobile manufacture, while firms such as Microsoft or Intel make much more money supplying all the computer manufacturers than they would as computer manufacturers themselves. Brief reference to eight LFs, albeit chosen in a non-random way, helps illustrate the variability and interdependent nature of LF strategy (Table 5.2). These eight firms, four based in North America and four based in Japan, are 'classic' LFs, at least as of the early 1990s. All of these firms are focused differentiators or component suppliers, targeting specific market niches within specific industries. In 1992, these companies had sales which ranged from around $300 million to over $1 billion. All were entrepreneurial, highly innovative, had implemented extensive marketing networks and were committed to ongoing R&D programs. In the case of Cisco Systems, a Silicon Valley-based manufacturer of internetworking gear for computers, with 1996 sales of over $4 billion, the R&D budget was over $300 million while expenditures on its global 'two-tier' distribution system were over $700 million. In the more modest case of Invacare, a Cleveland-based manufacturer of wheelchairs and related home care medical products, 1994 expenditures on R&D amounted to $7 million on sales of $411 million and it too had evolved an extensive marketing system featuring close contact with independent home health care dealers.

Each of the eight cases can be classified as a market leader (in a specialised market), component supplier or competitor (to giant end-use firms). A separate category is designated for leader firms, that is firms which lead an industry not dominated by MNCs. For example, Invacare is the overall leader in a market which limits it to LF size, but its ambitions are such that it has publicly stated it wants to be a giant, included in the *Fortune* 500. However, leadership is a characteristic common to the other two categories. Among the component suppliers, the auto component suppliers, Magna and F-Tech, for example, are highly innovative 'leaders' in the development of particular components. F-Tech is a Tier 1 supplier of suspension systems for Honda, and its leadership focuses on designing a component that enhances the value of

Table 5.2 Business segmentation, industry and product type for selected LFs

Firm	Business segment	Industry or product	Product niches
Invacare	Leader	Medical-consumer	Wheelchairs, crutches
Dr. Pepper	Competitor	Beverages-consumer	Soft drinks
Cisco Systems	Leader	Computers-component	Internetworking gear
Magna	Component supplier	Autos-component	Diversified auto parts
F-Tech	Component supplier	Autos-component	Suspension systems
Murata	Leader	Machinery-capital	Textile machines, automated guided vehicles
Hirata	Leader	Machinery-capital	Factory systems
THK	Component supplier	Electronics-component	Custom-made bearings

Sources: For further details on these firms see Anderson and Holmes (1995) (Magna), Hayter (1997) (Invacare, Dr. Pepper and F-Tech), and Patchell (1993b) (Murata, Hirata and THK).

Honda's cars and manufacturing and procuring parts for that component. Magna, in contrast, is not an exclusive supplier and entertains the possibility of becoming an assembler competing with the car manufacturers (Anderson and Holmes, 1995, p. 666). Similarly, within the Japanese robot industry, Murata and Hirata are competitors with giant MNCs like Matsushita and Toshiba, but lead their market niche (Patchell, 1993b). Competition can of course be terminated, as witnessed by Dr. Pepper's recent takeover by Seven Up. On the other hand, maybe Cisco Systems, whose sales grew more than twelve-fold between 1992 and 1996 (from $340 million to over $4 billion) should now be considered a giant, although such a claim may be premature given its present profitability problems.

Indeed, it is the innovation-driven dynamism of LFs and their role as leaders in a smaller market, as competitors to MNCs, and as component suppliers that define their broadly-based strategic purpose or social rationale.

Internal Development of the Large Firm

The competitive advantages of LFs spring mainly from strengths developed internally through organisational and technological innovations. The source of these innovations, similar to SMEs, arises from the corporate priorities and cultures comprehensively shaped by dominating entrepreneurs. Thus in LFs, decision-making structures are typically flat and responsive, and employee relations are flexible and intimate. But in contrast to SMEs, the organisation achieves a higher level of sophistication and systemisation and, for example, R&D and marketing functions of LFs operate continuously and on a much bigger scale. Indeed, marketing and distribution systems are international in scope and typically supported by direct foreign investment in manufacturing facilities. Consequently, LFs are more powerful and diversified exploiters of economies of scale and scope than SMEs. In particular, for LFs, firm-level economies of scale (especially in R&D, marketing and those associated with multi-plant operations) are important and add to factory-level and product-level economies of scale. Similarly, the close integration of R&D, marketing and production, along with flexible labour, likely facilitate a richer vein of scope economies in LFs than is possible in SMEs. Finally, in contrast to SMEs, the fixed costs of R&D and marketing systems in LFs constitute a cause and effect of growth. For LFs, growth itself is much more likely to be an important motivation than is the case with SMEs, which typically operate within some self-imposed (if not market-imposed) size limit.

In relation to MNCs, as noted, LFs have similarly internationalised their operations and reaped substantial scale economies. In contrast to MNCs, LFs have an entrepreneurial culture, they are less bureaucratic and have developed around flexible labour practices. Indeed, Simon (1992, p. 122) suggests that labour problems in LFs 'absorb much less managerial energy' than in MNCs (Simon, 1992, p. 122). His rule-of-thumb based on German experience is that while management in large companies typically spend about 65 percent of their time dealing with bureaucratic and labour relations issues ('overcoming internal resistance') the figure is closer to 25 percent in LFs. Less restricted by these problems, LFs can concentrate on empowering and enskilling (flexible) workforces to achieve maximum creativity, cooperation and commitment to innovation. Moreover, if MNCs emphasise global mass marketing of a wider or full product range, LFs are global niche players and their innovation efforts are more specialised in terms of product-markets.

In terms of product-markets, and other respects, LFs distinguish themselves from SMEs by fuller exploitation of economies of scale and scope, and from MNCs by more focused orientation on particular niches. These distinctions also have implications for the function and organisation of factories which, at a minimum, are responsible for the execution of manufacturing processes. In this context, Fruin's (1992) concept of the 'focal factory', derived form Japanese experience, has broad relevance for the path to development taken by LFs. Thus, focal factories are designed to fully exploit product specialisms and all closely related diversification opportunities. In Fruin's terms, such factories are 'production site[s] with appended planning, design, development, and process-engineering capabilities, plus an ambition to accumulate, combine, and concentrate experience for the propagation and improvement of products and processes' (Fruin 1992, p. 24). Focal factories are complex, innovative and enjoy significant autonomy, intimately linking markets and production, defining localised 'architectures of innovation' which constantly search to exploit intra-firm economies of scope (Fruin, 1992, p. 212).

In highly innovative environments, which demand increasingly fast responses, focal factories are likely to become more important. Indeed, Kenney and Florida's (1993) underlining of the role of Japanese factories as laboratories is consistent with that of focal factories and in their view is an idea that is relevant elsewhere, including the US. Similarly, Cooke and Morgan's (1993) reference to 'fractal factories' hints of similar developments in Germany. Clearly, focal factories are not limited to LFs and are found within MNCs, as Fruin (1992) illustrates in the context of Toshiba. Indeed, the reorganisation of many MNCs from the strict hierarchical structures associated with Fordism that prescribed factories to the specialised role of 'basic work processes' (Hymer 1960), into more varied 'heterarchies' (Hedlund, 1986) and 'glocalised' structures (Swyngedouw, 1992, 1997), intimates a growing role for focal factories. Possibly, however, it is LFs that have provided the model, and competitive stimulus, for change among MNCs, rather than the other way around.

Spatial Connections

Strategic Base of the Large Firm

The geographical implications of LFs are not limited to employment or development concerns. In particular, the question arises as the extent to which the ability of LFs to 'thwart monopolies' encompasses a more equitable distribution of decision-making power that counters the geographic concentration of control activities favoured by MNCs. This proposition is not necessarily based on any explicit motives by LFs in favour of regional equity or power-sharing per se, but reflects tendencies for entrepreneurs, known to be significant among SMEs, to locate head-offices according to personal preferences, which for SMEs, typically means a 'home' location. Some indicative, if by no means systematic, evidence on head-office locations among LFs is provided by the head-offices of *Business Week*'s (1993) list of 50 US-based 'mid-sized giants'. In broad regional terms, the LF distribution is similar to the distribution of the leading (100) industrials from the *Fortune* 500 list (Hayter, 1997, pp. 211-212). The only regional outlier is a head-office in Boise, Idaho; none of *Fortune*'s (1996) top 100 industrials are in the mountain states. Otherwise, LF headquarters show a strong concentration around New York and the states of the Manufacturing Belt, plus secondary concentrations in Texas and California. Again bearing in mind the casual nature of the comparisons, *Fortune*'s top 100 also show slightly more preference for the south-east than *Business Week*'s LFs.

At intra-regional scales, however, the head-offices of *Business Week*'s (1993) LFs reveal more geographic variety than is evident among the *Fortune* 100 firms. Just ten firms have headquarters in metro centres, five are in inner suburbs, and 17 are in outer suburbs, while the remainder are in small rural towns, such as Paoli, Pennsylvania and intermediate-sized cities, such as Boise and Kalamazoo. This distribution may not imply regional or intra-regional equity but it does indicate wider location ranges of tolerance for control functions than is sometimes supposed, suggesting that even small places can benefit from the jobs and wealth of head-offices and have an impact on globalisation processes.

It is also worth noting that in a German context, Simon (1992, p. 122) made a special point of emphasising that:

... the typical 'hidden champion' is located in a small town or village rather than a big city. Few can be found in urban centres like Hamburg, Munich, or

Cologne, but many are scattered in places like Neutraubling, Harsewinkel, Tauberbischofscheim, Melle, and Stockdorf – towns most Germans have never heard of.

Similarly, in Japan, Nakamura's (1990) argument that LFs have been 'backbone' firms in the Japanese economy as a whole for the last 50 years also explains the development of many LFs outside the Tokaido (Tokyo-Osaka) core. In the robot industry, in particular, LFs were able to develop in remote rural areas as well as in the traditional core centres of Tokyo and Osaka (Patchell, 1993b). However, in some contrast to the US and Germany, there is a tendency of most regionally-based Japanese LFs to establish a Tokyo headquarters or even move headquarters to Tokyo for better access to markets and information.

Large Firm Factories and Local Relations

Much of the innovative strength of the LF is drawn from their entrepreneurial dimension and their focal-factory type organisation. Indeed, these characteristics imply local embeddedness, defined broadly in terms of localised political and cultural ties as well a economic linkages, at least with respect to home regions. Because LFs are powerful innovators and because they supply demanding high-quality market niches on a global basis, LFs are individually big enough to exert big employment, income and quantity and quality multiplier effects on local economies. Their impact is further enhanced because they are typically organised as learning systems. At the same time, the size of LFs and their global role typically stimulates factory location outside of their home base. Thus LFs are faced with resolving the tensions between locating factories to serve global markets most efficiently and the advantages of embeddedness and of maintaining the strength of focal factories. This tension, it might be noted, remains evident among MNCs and claims that giant firms have no particular geographical loyalties have been countered by studies that argue that MNCs have regional preferences and are embedded in home regions (Dicken, Forsgen and Malmberg, 1994; Hirst and Thompson, 1996). LFs define processes which help explain the connections between local embeddedness and internationalisation.

Resolving the tension between new factory location and maintaining innovative factories is one of the most creative forms of connectivity developed by LFs. In this regard, Mel Mixon and Frank Stonach, the entrepreneurs who own and direct Invacare and Magna, respectively, provide quintessential examples (see Table 5.2). Both entrepreneurs have the reputation as dominant, forceful personalities who have personally shaped the structure and strategies of their firms, including considerable DFI, while focusing decision making, R&D and important manufacturing functions in home regions (Hayter, 1997; Anderson and Holmes, 1995).

Embeddedness of course does not necessarily equate to local development. After all, in the original segmentation model, SMEs are defined as peripheral and frequent exploiters of low-wage labour. In the case of Magna, Anderson and Holmes (1995) argue that a key dimension of the firm's innovativeness is to combine high skill with low wages, notably by creating an 'industrial concept' of small specialised facilities each located in a different, if nearby community. Thus, Magna's founding entrepreneur has created a set of operating groups, coordinated by an executive management, which he heads. Each group of 10-20 plants comprises a geographically-concentrated network of small, specialised plants which focus on particular modular sub-assemblies, a strategy which Magna has helped pioneer in North America (Anderson and Holmes, 1995, p. 664). Most plants are highly automated, flexible and have less than 100 employees. Given the skilled workers available in southern Ontario, Anderson and Holmes argue that Magna's structure allows the firm to enjoy lower wages because labour power is limited at each location; the firm has the possibility of playing off one location against another while limiting the extent of capital investment in any one location. At the same time, each group has its own R&D and engineering staff, an ability to develop its own capital goods, marketing responsibilities and human resources management, while each plant is a profit centre, has a profit-sharing scheme and is responsible for training, tooling and the identification of markets and product development. Magna also provides social and recreational facilities. Moreover, skill is a problematical concept, as is the definition of low wages. In the manufacturing sector, it is unlikely that high wages can be achieved and sustained in the absence of innovation.

Large Firm Production Systems in Space

For LFs, the options for organising patterns of production, within factories in the firm and among networks of firms, in particular localities to fulfill global strategies are considerable. From the point of view of local development, a key issue is the extent to which LFs externalise or internalise learning (Figure 5.2). LFs, even if relatively specialised, are big enough to comprise complex value

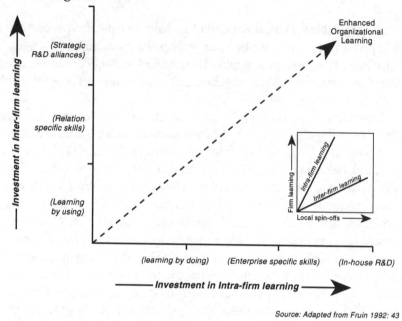

Source: Adapted from Fruin 1992: 43

Figure 5.2   Learning within and among firms and local spin-offs

Source: Adapted from Fruin (1992, p. 43).

links or chains which can be allocated within an internal division of labour, externally distributed among a social division of labour, or some combination of both. It is increasingly recognised, however, that the transaction-cost basis for discriminating between internal and social divisions of labour is complex, even ambiguous, and influenced by particular historical, cultural and entrepreneurial considerations (Odaka *et al.*, 1988). Indeed, arguments can made that both internal and social divisions of labour facilitate innovation of high-quality products by LFs. Thus, the attraction of the social division of labour is to provide access to the scale and scope economies of flexible specialisation, based around the vitality of a diverse range of entrepreneurial inputs as well as to labour pools of varying cost and skill. The attraction of the internal division of labour, on the other hand, is to protect proprietary knowledge and to allow direct control over crucial links which firms may judge important from the point of view of quality control, quality enhancement and security.

In Germany, for example, Simon (1992) claims that one of five 'common practices' defining the competitive advantages of Germany's hidden champions is 'considerable self-reliance on technical competence' (the others being product market specialisation with geographic diversity; an emphasis on customer value; an emphasis on blending technology with customer needs; and mutual interdependence between managers and employees). Despite Germany's own long-standing subcontracting traditions, the clear implication of this view is that LFs believe that learning and bargaining can be best organised internally to meet long-run motivations. Indeed, according to Simon (1992, p. 122), reliance on internal technical know-how is bolstered by their location isolation in remote, small communities. Such a setting, it is argued, not only allows for cooperative labour relations but also limits loss of know-how to outsiders. In a North American context, Magna, Invacare and Cisco Systems, despite radically different geographical configurations, are also organised around internal divisions of labour and control of know-how. In addition, while Cisco Systems is concentrated in Silicon Valley, Invacare, like Magna, has often located branch plants in peripheral, rural regions.

More research needs to be done to clarify production system linkages of German and American LFs. These linkages may well be substantial. It is likely that even the most innovative LF will rely on other firms to supply it with materials, components, parts, and services. Indeed, what is important is the selectivity and nurturing of strategic supplier and collaborative arrangements. The fact that component suppliers like F-Tech or Magna are themselves part of production systems indicates the wisdom of core firms working with innovative suppliers and hence it is unlikely that the entrepreneurs of LFs will ignore outsourcing advantages. Through those connections LFs receive great amounts of technological, market, and managerial know-how that they utilise not only in relation with the original customer, but which they can convert to develop new efficiencies, products, and business with other customers.

In Japan the development of production systems with high levels of information exchange is a strong source of competitive advantage. As is now well known, there is a profound commitment to the social division of labour and, to a significantly greater degree than elsewhere, firms in Japan realise motives related to profits, efficiency, control and security by externalising learning and bargaining processes (Asanuma, 1985a,1985b, 1989; Dore, 1986; Nishiguchi 1994; Odaka *et al.*, 1988; Patchell, 1993a, 1993b; Smitka, 1991). While some observers see the roots of this tendency in terms of culture and Confucianism, Fruin (1992) stresses another kind of historical argument: that, given Japan's position as a late developer, the only way Japanese firms could effectively learn, and ultimately improve upon, western technology was by specialising and depending upon one another. While traditionally, inter-firm relations were primarily seen in the exploitative terms of the original dual economy model, there is increasing emphasis on cooperative, mutually beneficial relations (Asanuma, 1989; Fruin, 1992). In either case, inter-firm relations have been depicted primarily in terms of relations between core firms and SMEs (while SME- dominated production systems are also recognised). These core firms include LFs, who commonly have cooperative firm associations that compare to those of MNCs in levels of intense information interaction between firms. They do not have the same levels of personnel or investment exchange, however. Across the board, Japanese leader, competitor, and component supplier LFs rely on external production systems to supply them with scale and scope economies that they cannot achieve in-house. Indeed, these economies make their position within the social division of labour possible. As the Japanese robot industry reveals, reliance on outsourcing also makes the existence of competitors viable.

The Japanese Robot Industry

The Japanese have developed the world's largest robot industry; indeed Japan dominates this industry. That domination involves the efforts of many firms. Sales in the industry are led by the divisions of MNCs such as Matsushita (134,000 employees), which has a robot division, or spin-offs of MNCs such as Fanuc which originated within Fujitsu. They are accompanied by Ishikawa Harijima, Mitsubishi, Hitachi, and others. In addition to these multi-divisional MNCs, LFs such as Yaskawa, Hirata, Murata, and several others have captured important niche markets in the industry. Research conducted on four firms in the industry was designed to specifically reveal the characteristics of firms in these different segments (Patchell, 1991; 1993a, 1993b). Matsushita, in addition to being the world's largest robot manufacturer is one of its largest electronics manufacturers and is an MNC and 'leader'. Yaskawa, Hirata, and Murata have their own distinctions, and may be classified as LFs that are leaders and competitors (Table 5.3).

Segment Connectivity and Industrial Organisation: The Role of Large Firms

Whether or not LFs are actually or potentially threats to MNCs is only part of their significance – most important is their influence on the competitive vitality

Table 5.3   Giants and loyal opposition: the Japanese robot industry

Company	1988 sales (¥100 million)	Robot specialisation
Giant firm		
Matsushita	816	Material handling, especially Cartesian robots, 50 percent for external markets
Loyal opposition		
Yaskawa	180	Multi-articulated playback machines for cutting, grinding, and welding (especially for auto industry); robots incorporate its leading-edge motion control technology
Hirata	32	Most robots (90 percent) sold as part of manufacturing systems; arm base and machine base, especially Cartesian robots designed for assembly purposes
Murata	125	Automated guided vehicles which incorporate robots and sold to many industries

Source: Based on Patchell (1991). This source lists all important robot manufacturers in Japan.

of an industry. In the following example of the Japanese robot industry we see the vital role that LFs play as competition to the major manufacturers and as component suppliers. In addition to those three segments of the industry, each of these firms has its own supplier network. However, the vitality of the the industry is not revealed by describing its segments, but by analysing how direct and indirect inter-firm learning occurs (supplemented by substantial intra-firm learning). The connectivity between giant manufacturer, LF competitor, LF component supplier and parts supplier *and* connectivity of all competitors through their common component suppliers explains the vitality of the Japanese robot industry.

Matsushita's robot production has evolved from constant improvement of its production processes. The Corporate Production Engineering Division undertakes these process improvements and develops machines for internal use and external sales (Patchell, 1991, p. 631). The most important machines made by Matsushita are related to electronics assembly (e.g. electronic component insertion machines and Cartesian robots), but as the firm wants to be a full-range robot supplier it purchases heavy-duty machines from ABB. These combinations allow Matsushita to supply complete automated systems to a range of customers.

Like Matsushita, Yaskawa is multi-divisional firm with a long history, but with 4,100 employees it has always focused on a much smaller product range. Yaskawa's robot production is a relatively new division in the firm which grew as a electric motor company supplying Japan's mining, steel-making, and other heavy industries. With the relative decline of these industries in the 1970s Yaskawa chose to pursue a new role in producing robots aimed at the automobile market (Patchell, 1991, p. 576). The firm sought to develop market niches by innovating highly sophisticated robots used in welding, cutting, and grinding operations which incorporate its established expertise in motor-control technologies. Yaskawa's arc welding robots quickly dominated the world market, while the firm has developed expertise in the manufacture of peripheral devices used in robots and in related software packages. Robot equipment is now almost as important as electrical machinery to the sales of the firm.

Hirata (1,020 employees) and Murata (2,900 employees) are smaller, newer, entrepreneurial firms with precise niches in the robot industry. Hirata produces complete manufacturing (primarily assembly) systems to order for many of the world's top manufacturing firms. It also produces robots for its manufacturing systems and for external sale. The strength of Hirata's expertise in this area is such that some of Japan's largest electronics companies – which make their own robots – contract with it to make their assembly lines. Hirata is also unique for its location in the small city of Kumamoto, distant from the primary manufacturing centres and not known for its manufacturing base. The success of this firm is due to the vision of its founder and his willingness to work with and foster local suppliers. Murata is a second-generation firm whose original and primary business is textile machinery manufacture. The second-generation president has led the firm into a limited and related diversification. One of these is the production of automated warehouse systems and it makes the automated guided vehicles critical to their function. Murata is one of the top three manufacturers of these systems.

A large number of firms can enter the robot industry and succeed among the giants because all the robot manufacturers are dependent on highly specialised component manufacturers. These components include motors, pneumatic devices, transmissions, and bearings. Firms that specialise in these components can gain economies of scale not possible even in the largest MNCs and, through their efficiency advantages, often gain majority market shares. Because of the benefits of their products any robot maker wishing to produce a competitive product must use many of the same firm components as their competition; thus the component makers are often in a more favourable position than the robot makers themselves.

Three of these specialist firms are THK, SMC, and Harmonic Drive. Qualitatively, they are similar in that they produce components essential to most robot makers, but they exemplify how LFs may differ in size despite the fact that they can play a key role in an industry. THK is a producer of the linear bearings crucial to the advancement of the robot industry and employs about 1,800. The company's success is built upon the linear bearing invented and patented by the founder Teramachi. Because of the requirements for constant intense and dependable performance in the robots and machine tools which are THK's market, the machining and precision of the bearings has to be performed to an extreme of level of tolerance. Thus, THK has invested considerably in specialised and integrated factories. Teramachi provides strong, 'hands-on' leadership. He has been particularly emphatic about installing engineering know-how in the workforce and among sales staff to facilitate communication between consumers and the firm. THK pays constant attention to innovation and is an acknowledged world leader in ball bearing systems, selling its technology globally and taking out many patents in the US, Europe, and Japan.

SMC is somewhat larger than THK with 2,600 full-time and 1,400 part-time employees. Its specialty is pneumatic devices, originally developed from a sintering process, to help articulate the uses of robots. Like THK, the firm is entrepreneurial, but rather than focus on one specific product technology and the robot/machine tool industry, SMC chooses to diversify the scope of pneumatic devices and industries across which it sells. To diversify, it uses customer input to develop new products and standardise them for global markets. It has about 10 percent of the world market for pneumatic devices with 16 subsidiaries or joint ventures and plants in Australia, Asia, North America, and Europe, and over 30 offices worldwide.

Harmonic Drive is a much smaller firm (260 employees), but still a powerful entity in the robot industry because the electric motor transmission that it makes is indispensable for many of Japan's robot makers. Again the company's main technology was developed by its president (in collaboration with an American company) and the company is run by him. Although the demand and varieties for the device are limited by the number of motors used in robots (the industry accounts for 60 percent of production), the company uses information from customers to further develop its product range.

Finally, a significant feature of segmentation in the robot industry is the important role played by SMEs. MNCs like Matsushita, loyal opposition, specialist LFs like Hirata, Yaskawa, and Murata, and LF component specialists use, and indeed are dependent on, SME specialist parts suppliers and processing firms. These firms are almost always organised into cooperative firm associations, although the level of formality and intensity of interactions of core firm and supplier vary significantly from one association to another. The outstanding aspect of these relationships is that both diversified manufacturers, such as Matsushita, and focused specialists, like THK, find benefits in outsourcing an average 50 percent of production. For Matsushita, outsourcing has allowed concentration on new products and, for all firms, outsourcing allows maintenance of core competencies. Many of the SMEs interviewed (Patchell, 1991) were also developing integrated production systems and their entrepreneurs seemed capable of making the organisational step to LF status. Most often, the lacking ingredient was a proprietary technology which would make this step feasible. However, in this regard their LF relationship becomes all the more important; the LFs actively foster the development of these firms. Thus in addition to their direct impact on local economies, LFs have an even greater indirect effect on the upgrading of localised firm networks through inter-firm learning.

Learning and Bargaining Within the Social Division of Labour

The social divisions of labour organised by the robot manufacturers (the giants and LFs) and component specialists – with at least the component specialists crossing firm-based systems – fundamentally evolve as learning systems, as defined by the relation specific skill. On the one hand, core firms dominate the exchange of technological expertise in exchanges with *design supply* (DS) suppliers, while *design approved* (DA) suppliers are those that have expertise the core firm does not. Whereas core firms provide the basis for DS suppliers to learn (in the form of demonstrations, blueprints and skills), DA suppliers use their know-how to continually respond to the core firm's specific needs (specifications). In the middle of the DS-DA continuum of possibilities, core firms and suppliers exchange know-how on more or less equal terms. Moreover, core firms reward the innovativeness of DA suppliers by allowing greater profit margins in price negotiations, increasing or at least stable markets (wherever possible), and increasingly sophisticated challenges. In addition, DA suppliers have the know-how to diversify markets and products. In other words, there are incentives for suppliers to shift from DS to DA status.

Certainly, the relative bargaining power of the robot manufacturers (and component specialists), rooted in their expertise, control over assembly and marketing networks, is evident in relations with suppliers, especially through the two-vendor policy and pricing contracts, and tender policy (and related quantity contracts). Thus the two-vendor policy reflects the tendency of core firms (contractors) to have at least two suppliers for a particular category of good. Tenders for component contracts for new models keep suppliers in competition with one another and some firms demand price reductions every six months by suppliers based on learning curve economies. By these initiatives, core firms place pressure on suppliers and maintain 'face you can see' competition. Yet, these same initiatives facilitate innovative and cooperative behavior (Ito, 1994). Thus, innovative DA suppliers are rewarded by waiving price reductions and all worthy suppliers are awarded business, while competition among suppliers is oriented towards product quality as well as cost. In the anatomy of the Japanese robot production system, bargaining and learning networks are closely intertwined, with the former designed to support the latter (Patchell, 1993a, 1993b).

In practice, the suppliers in the robot industry play vital roles in innovating products which enhance performance and reduce costs. Some DA suppliers, as well as component specialists, enjoy levels of expertise equal to or surpassing that of the core firms and may even create products not conceived by the robot manufacturers. Even exclusive DA suppliers can achieve the size and characteristics associated with LFs, and they may be encouraged by core firms to seek other markets. In other words, in the robot industry subcontractors are frequently dynamic innovators. Specialist component suppliers such as THK are highly independent, capable of pursuing global, focused market strategies with considerable bargaining power in their own right. Moreover, the component specialist LFs that cut across core-firm based production systems add to the impressive flow of information among firms in the robot industry. While such non-exclusive links to important suppliers may be a source of tension when occurring on a specific product basis (as was the case with Hirata's technological cooperation with THK), they nevertheless help bolster inter-firm learning.

LFs therefore play vital roles in the Japanese robot production system, as counter magnets to the giants, as 'core' firms exploiting all market niches, and as creators of externalised production systems in which subcontractors can be pro-active. LFs in this industry have also geographically extended the robot system in Japan. Within the Tokyo-Osaka (Tokaido) industrial axis are Matsushita, based in Osaka, and Murata, headquartered in Kyoto, but Yaskawa and Hirata developed their robot systems in Kyushu. Yaskawa, located in the former steel town of Yahata, and Hirata, located in Kumamoto, an agricultural area in the centre of Kyushu with no tradition of machinery manufacture, are even more remote from the core region.

Yet, Hirata has not rationalised location isolation with the need for secrecy and dependency on intra-firm learning. Rather, it has developed an extensive supplier system in the region (Table 5.4). Thus Hirata has developed a localised vertical division of labour within Kyushu and, while it is connected to LFs headquartered in other regions of Japan as part of an horizontal division of labour, several of these firms have branch plants in Kyushu. DA suppliers are less important than in the case of Matsushita, for example, and DA and DS (and catalogue good) parts are purchased by two departments for the firm as a whole rather than by individual factories. Nevertheless, Hirata is assessing the possibilities of formally establishing a cooperative association (of suppliers). Moreover, Hirata's competitive edge is based on delivering one-of-a-kind orders for complete production lines in a short time, a strategy that requires subcontractors to meet delivery times which are often only one to three days (Patchell, 1993b, p. 933).

In terms of business segmentation, the Japanese robot industry illustrates that power networks and learning networks are not mutually exclusive, and that effective inter-firm learning requires appropriate bargains. Indeed, the relation-specific skill explicitly reveals the interdependence of learning and power networks. These networks are clearly non-hierarchical, not only in a business sense but also geographically. Indeed, in developing an intricate horizontal and vertical division of labour among its FFPS, the robot industry reveals the potential of peripheries in Japan not only as spaces for branch plants but as places where flexibly specialised production featuring high levels of inter-firm learning can develop. How such potential translates elsewhere (outside of Japan) is another matter.

Conclusion

LFs define a segment that has been hitherto unappreciated and reveal segmentation to be a more dynamic process than hitherto presented. The distinctive contribution of LFs lies in their special role as innovators. From this perspective, the recognition of LFs goes beyond adding another layer of segmentation. Rather, LFs highlight the role of strategic connectivities in the relationships between business organisation, innovation and local development. Thus, LFs connect SMEs with MNCs, they connect the local with the global, and they connect learning and power networks. These connections are strategic and inter-related. LFs are reminders that dualisms are abstractions and that the reality of segmentation needs to be juxtaposed with the reality of interdependence, or connections. Moreover, explicit recognition of LFs invites interpretation of segmentation and connections as processes, rather than simply static structures. In particular, LFs underline the vital role of innovation in generating (and destroying) segmentation, and connections within and among firms, both with respect to learning as well as power relations.

A consideration of LFs has implications for local and regional development policies that have evolved within the confines of the dualism framework. Historically, since the 1950s, regional policies first favoured the 'external' attraction of branch plants controlled by MNCs and in more recent years greater emphasis has been placed on the 'internal' promotion of SMEs. Many contemporary policy packages seek to do both. Yet policies that focus on MNCs and SMEs as separate segments ignore the complexity of the segmentation processes that exist within the size distribution of firms, and the nature of the ties between segments, ties that include issues of learning as well as of bargaining. Local policy needs to look more comprehensively at populations of firms and the ways in which they compete or complement one another. It may be for example, that local economic development is not just a question of missing SMEs, or branch plants, so much as backbones.

Acknowledgements

We are grateful for the comments of Neil Alderman and for the observations of Ed Malecki and Päivi Oinas. R. Hayter also thanks the SSHRC for financial support.

Table 5.4   Hirata's Kyushu-based production system

A. Hirata: Firms of horizontal division of labour

Company	Main business	Employment (circa 1990)	Relevant locations
Hirata Machinery Company	Manufacturing systems	1,020	Kumamoto Kyushu (HQ, MP)
Daiden	Wire and cable	500	Kurume, Fukuoka
Harmonic Drive	Electric motor transmissions	260	Tokyo (HQ); Hakata, Fukuoka (BO)
Yaskawa	Electric motors	399	Kokura, Fukuoka (MP); Kumamoto (BO)
THK	Bearings	1,780	Tokyo (HQ); Hakata, Fukuoka (BO)
Mitsubishi Nagasaki Machinery	Structural steel; industrial machinery; plant equipment	430	Nagasaki (HQ, MP)
Mitsubishi Electric Equipment	Electric equipment sales	31	Kumamoto (BO)
SMC	Pneumatic devices	2,505	Tokyo (HQ) Kumamoto (BO)
Omuta	Printed circuit boards	55	Omuta, Fukuoka (HQ, MP)

B. Hirata: Firms of vertical division of labour

Company	Main business	Employment (circa 1990.)	Relevant locations
Ogata	Plating and surface treatment	100–110	Kumamoto (HQ, MP)
Yoshino	Machinery	40	Kumamoto
Kashu	Steel cutting	29	Yahata, Fukuoka
Nishioka	Sheet metal fabrication	16	Kumamoto
Daiichi	Press fabrication	23	Yahata, Fukuoka
Tsuchiya Gum	Plastic and rubber parts	100	Kumamoto
Terano	Metal machining	24	Kumamoto
Kumamoto Spring	Spring manufacture	10	Kumamoto
Daiich Tekko	Structural steel	13	Kumamoto
Ishihara Harness	Wire harnesses	30	Kumamoto
Kawano	Steel structures	6	Kasuya, Fukuka
Kaseda	LCD and LED devices	100	Kaseda; Kagoshima
Togo	Metal machining	10	Kagoshima
Senko Screen	Printing	7	Kumamoto

*(MP) main plant, (HQ) headquarters, (BO) transaction branch office
Source: Patchell (1991, p. 475).

References

Amin, A. and Robins, K. (1991), 'These Are Not Marshallian Times', in R. Camagni (ed.) *Innovation Networks: Spatial Perspectives*, Belhaven Press, London, pp. 105-118.

Anderson, M. and Holmes, J. (1995), 'High-Skill, Low-Wage Manufacturing in North America: A Case Study from the Automotive Parts industry', *Regional Studies*, vol. 29, pp. 655-671.

Asanuma, B. (1985a), 'The Organization of Parts Purchases in the Japanese Automotive Industry', *Japanese Economic Studies*, vol. 13, pp. 32-53.

Asanuma, B. (1985b), 'The Contractual Framework for Parts Supply in the Japanese Automotive Industry', *Japanese Economic Studies*, vol. 13, pp. 54-78.

Asanuma, B. (1989), 'Manufacturer-Supplier Relationships in Japan and the Concept of the Relation-Specific Skill', *Journal of the Japanese and International Economies*, vol. 3, pp. 1-30.

Averitt , R.T. (1968), *The Dual Economy: The Dynamics of American Industry*, Norton, New York.

Berle, A. and Means, G. (1932), *The Modern Corporation and Private Property*, Macmillan, New York.

Britton, J.N.H. (1991), 'Reconsidering Innovation Policy for Small and Medium Sized Enterprises: The Canadian Case', *Environment and Planning C: Government and Policy*, vol. 8, pp. 189-206.

Business Week (1993), 'The Little Giants', 6 September, pp. 40-46.

Cooke, P. and Morgan, K. (1993), 'The Network Paradigm: New Departures in Corporate and Regional Development', *Environment and Planning D: Society and Space*, vol. 11, pp. 543-564.

Dicken, P., Forsgren, M. and Malmberg, A. (1994), 'The Local Embeddedness of Transnational Corporations', in A. Amin and N. Thrift (eds), *Globalization, Institutions, and Regional Development in Europe*, Oxford University Press, Oxford, pp. 23-45.

Dore, R.P. (1986), *Flexible Rigidities: Industrial Policy and Structural Adjustment in the Japanese Economy, 1970-80*, Stanford University Press, Stanford.

Fortune (1996), 'The 500 Ranked within States', April 26, pp. F31-42.

Fruin, W.M. (1992), *The Japanese Enterprise System*, Clarendon Press, Oxford.

Galbraith, J.K. (1952), *American Capitalism*, Houghton Mifflin, Boston.

Galbraith, J.K. (1967), *The New Industrial Estate*, Houghton Mifflin, Boston.

Glasmeier, A. (1988) 'Factors Governing the Development of High-Tech Industry Agglomerations: A Tale of Three Cities', *Regional Studies*, vol. 22, pp. 287-301.

Hayter, R. (1997), *The Dynamics of Industrial Location: The Factory, the Firm and the Production System*, Wiley, Chichester.

Hayter, R. and Edgington, D. (1997), 'Cutting against the Grain: A Case Study of MacMillan Bloedel's Japan Strategy', *Economic Geography*, vol. 73, pp. 185-210

Hayter, R. and Watts, H.D. (1983), 'The Geography of Enterprise: A Reappraisal', *Progress in Human Geography*, vol. 7, pp. 157-181.

Hedlund, G. (1986), 'The Hypermodern MNC – A Heterarchy?' *Human Resource Management*, vol. 25: 9-35.

Hirst, P. and Thompson, G. (1992), *Globalization in Question*, Polity Press, Cambridge.

Holmes, J. (1986), 'The Organization and Locational Structure of Production Subcontracting', in A. J. Scott and M. Storper (eds), *Production, Work, Territory*, Allen and Unwin, Boston, pp. 80-106.

Hymer, S.H. (1960) *The International Operations of National Firms: A Study of Direct Investment*, Ph.D. dissertation, Massachusetts Institute of Technology, Cambridge, MA.

Ito, M. (1994), 'Interfirm Relations and Long-Term Continuous Trading', in K. Imai and R. Komiya (eds), *Business Enterprise in Japan: Views of Leading Japanese Economists*, MIT Press, Cambridge, MA, pp. 105-115.

Kenney, M. and Florida, R. (1993), *Beyond Mass Production*, Oxford University Press, Oxford.

Kuhn, R.L. (1982), *Mid-Sized Firms: Success Strategies and Methodologies*, Praeger, New York.

Kuhn, R.L. (1985),*To Flourish Among Giants*, John Wiley, London.

Nakamura, H. (1990), *New Mid-sized Firm Theory*, Tokyo Keizai Shinpansha, Tokyo (in Japanese).

Nishiguchi, T. (1994), *Strategic Industrial Sourcing: The Japanese Advantage*, Oxford University Press, New York.

Odaka, K., Ono, K., and Adachi, F. (1988), *The Automobile Industry in Japan: A Study of Ancillary Firm Development*, Kinokuniya, Tokyo.

Oinas, P. (1995), 'Types of Enterprise and Local Relations', in B. van der Knaap and R.B. Le Heron (eds), *Human Resources and Industrial Spaces: A Perspective on Localization and Globalization*, John Wiley, London, pp. 177-195.

Patchell, J. (1991), *The Creation of Production Systems within the Social Division of Labour of the Japanese Robotics Industry: The Impact of the Relation Specific Skill*, Unpublished Ph.D. thesis, Simon Fraser University, Burnaby.

Patchell, J. (1993a), 'From Production Systems to Learning Systems: Lessons from Japan', *Environment and Planning A*, vol. 25, pp. 797-815.

Patchell, J. (1993b), 'Composing Robot Production Systems: Japan as a Flexible Manufacturing System', *Environment and Planning A*, vol. 25, pp. 923-944.

Patchell, J. (1996), 'Kaleidoscope Economies: The Processes of Cooperation, Competition and Control', *Annals of the Association of American Geographers*, vol. 86, pp. 481-506.

Patchell, J. and Hayter, R. (1995), 'Skill Formation and Japanese Production Systems', *Tijdschrift voor Economische en Sociale Geografie*, vol. 86, pp. 339-356.

Piore, M. and Sabel, C.F. (1984), *The Second Industrial Divide: Possibilities for Prosperity*, Basic Books, New York.

Porter, M.E. (1985), *Competitive Advantage*, The Free Press, New York.

Porter, M.E. (1990), *The Competitive Advantage of Nations*, Free Press, New York.

Robertson, D.H. (1928), *The Control of Industry*, Nisbet, London.

Simon, H. (1992), 'Lessons from Germany's Midsized Giants', *Harvard Business Review*, vol. 70, 2, pp. 115-123.

Smitka, M.J. (1991), *Competitive Ties: Subcontracting in the Japanese Automotive Industry*, Columbia University Press, New York.

Steed, G.P.F. (1982), *Threshold Firms: Backing Canada's Winners*, Science Council of Canada, Background Study 48, Ministry of Supply and Services, Ottawa.

Storper, M. (1993), 'Regional "Worlds" of Production: Learning and Innovation in the Technology Districts of France, Italy and the USA', *Regional Studies*, vol. 27, pp. 433-455.

Swyngedouw,E.A. (1992), 'The Mammon Quest: "Glocalisation", Interspatial Competition and the New Monetary Order: The Construction of New Scales', in M. Dunford and G. Kaflakas (eds), *Cities and Regions in the New Europe*, Belhaven, London, pp. 39-67.

Swyngedouw, E. (1997), 'Neither Global nor Local: "Glocalization" and the Politics of Scale', in K.R. Cox (ed.), *Spaces of Globalization*, Guilford, New York, pp. 137-166.

Taylor, M.J. and Thrift, N..J. (1982a), 'Industrial Linkage and the Segmented Economy: 1. Some Theoretical Proposals', *Environment and Planning A*, vol. 14, pp. 1601-1613.

Taylor, M.J. and Thrift, N.J. (1982b), 'Industrial Linkage and the Segmented Economy: 2. An Empirical Interpretation', *Environment and Planning A*, vol. 14, pp. 1615-1632.

Taylor, M. and Thrift, N. (1983), 'Business Organization, Segmentation and Location', *Regional Studies*, vol. 17, pp. 445-465.

Veblen, T. (1932), *The Theory of Business Enterprise*, Mentor Books, New York.

PART II
BECOMING CONNECTED

6 Local Learning and Interactive Innovation Networks in a Global Economy

Bjørn T. Asheim and Philip Cooke

Introduction

The focus in this chapter is on the evolution of innovation intervention, as policy-makers learn how to strengthen the localised innovation base through developing regional potential for systemic innovation. One of the main lessons learned from the rapid growth of industrial districts and other specialised areas of production was the understanding of industrialisation as a territorial process, underlining the importance of agglomeration and non-economic factors (i.e. culture, norms and institutions) for the economic performance of regions. Another lesson relates to the new view on innovation as a social process. This broader understanding of innovation as a social, non-linear and interactive process has emphasised the role played by socio-cultural structures in regional development as necessary prerequisites for regions in order to be innovative and competitive in a post-Fordist global economy. According to Amin and Thrift, this forces a re-evaluation of 'the significance of territoriality in economic globalisation' (Amin and Thrift, 1995, p. 8). The combined effect of understanding industrialisation as a territorial process, and innovation as a social process, has also dramatically changed the basis for launching industrial and technology policies.

Based on modern innovation theory it could be argued that firms in territorial agglomerations can develop their competitive advantage based on innovative activity, which is a result of socially and territorially embedded, interactive learning processes. At the same time, this view expands the range of branches that could be viewed as innovative – from typical high-tech branches of Silicon Valley to traditional, non R&D-intensive branches of peripheral regions.

Looking at the efforts by policy-makers to model high-tech innovation based on the examples of successful areas like Silicon Valley, most attempts have involved the idea of co-locating research centres and innovation-intensive firms in science and technology parks. In some cases this has involved designating whole cities as 'science cities' or 'technopoles'. Although benefits have been accrued from such plans, the literature also cites a frequent sense of disappointment that more has not been achieved. In cases drawn from France and Japan – countries that have arguably proceeded furthest with the *technopolis* policy – an absence of synergies has been observed among co-located laboratories and firms. Science parks, in themselves, have not always met the expectations of their founders and efforts have been made to learn from early mistakes. More recently, more attention is paid to the factors which lead to embeddedness (Granovetter, 1985) amongst firms and innovation support organisations. In the context of the present discussion, this idea can be seen as representing the institutional and organisational features of community and solidarity, and the exercise of 'social capital' (Putnam, 1993) as well as the foundations of high-trust, networked types of relationship among firms and organisations. It has been recognised to some extent that science parks are a valuable element but not the only or main objective of a localised or regionalised innovation strategy. Research has pointed at the nature and range of interaction among firms and organisations engaged in innovation (see, for example, Edquist, 1997; Braczyk *et al.*, 1998; Cooke and Morgan, 1998), and policy is moving towards a notion of the *region* as an important level at which strategic innovation support is appropriate (Asheim, 1996; Asheim and Isaksen, 1997; Tödtling and Sedlacek, 1997).

The first part of the chapter reviews the theoretical foundations of the regionalisation approach, and especially the interactive innovation network approach and the most recent thinking about the design of regional innovation systems. The second part examines examples of interactive-model innovation networks in which, to some extent, learning gains from observations of weaknesses of linear-model approaches were integrated into the design of more networked solutions. Conclusions will then be drawn concerning the key elements now considered essential to the optimal functioning of innovation support in order to promote endogenous regional development.

Theoretical Foundations of the Regionalisation Approach

Innovation as an Interactive Learning Process

Modern innovation theory stems from a criticism directed at the linear model of innovation which has guided the formulation of national R&D policies in the past. In Smith's (1994) view, the problem of this model is two-dimensional. The first problem was 'an overemphasis on research (especially basic scientific research) as the source of new technologies' (Smith, 1994, p. 2). Within this perspective a low innovative capacity could be explained by a low level of R&D activity. Consequently, technology policy in most western countries was directed towards increasing the level of basic research. The second problem was a 'technocratic view of innovation as a purely technical act: the production of a new technical device' (Smith, 1994, p. 2).

This criticism implies a broader view on the process of innovation as a technical *and* a social process; it involves interaction between firms and their environments and it is non-linear (Smith, 1994). In line with this criticism, the presently emerging innovation theory implies a more sociological approach to the process of innovation in which interactive learning is looked upon as 'a fundamental aspect of the process of innovation' (Lundvall, 1993, p. 61). Lundvall emphasises that 'learning is predominately an interactive and, therefore, a socially embedded process which cannot be understood without taking into consideration its institutional and cultural context' (Lundvall, 1992, p. 1). Also, Camagni emphasises that:

... technological innovation ... is increasingly a product of social innovation, a process happening both at the intra-regional level in the form of collective learning processes, and through inter-regional linkages facilitating the firm's access to different, though localised, innovation capabilities (Camagni, 1991, p. 8).

This alternative model could be referred to as a bottom-up interactive innovation model (Asheim and Isaksen, 1997). Thus, it is particularly appropriate for SMEs in networks and the 'learning economy', and greatly facilitated by geographical proximity and regional agglomerations. The interactive innovation model puts emphasis on:

... the plurality of types of production systems and of innovation (science and engineering is only relevant to some sectors), 'small' processes of economic coordination, informal practices as well as formal institutions, and incremental as well as large-scale innovation and adjustment (Storper and Scott, 1995, p. 519).

The rapid economic development in the 'Third Italy', based on territorially agglomerated SMEs in industrial districts (ID), has drawn an increased attention towards the importance of cooperation between firms and between firms and local authorities in achieving international competitiveness. Pyke (1994) underlines the close inter-firm cooperation and existence of a supporting institutional infrastructure at the regional level (e.g. centres of real services) as the main factors explaining the success of Emilia-Romagna in the 'Third Italy'. According to Dei Ottati,

> ... this willingness to cooperate is indispensable to the realization of innovation in the ID which, due to the division of labour among firms, takes on the characteristics of a collective process. Thus, for the economic dynamism of the district and for the competitiveness of its firms, they must be innovative but, at the same time, these firms cannot be innovative in any other way than by cooperating among themselves (Dei Ottati, 1994, p. 474).

Thus, if these observations are correct, they represent new 'forces' in the promotion of technological development in capitalist economies, implying a modification of the overall importance of competition between individual firms. Relying on Porter's (1990) empirical evidence, Lazonick argues, that 'domestic cooperation rather than domestic competition is the key determinant of global competitive advantage. For a domestic industry to attain and sustain global competitive advantage requires continuous innovation, which in turn requires domestic cooperation' (Lazonick, 1993, p. 4). Cooke (1994) supports this view, emphasising that 'the co-operative approach is not infrequently the only solution to intractable problems posed by globalisation, lean production or flexibilisation' (Cooke, 1994, p. 32).

This perspective emphasises the importance of organisational innovations to promote cooperation, primarily through the formation of dynamic flexible learning organisations. Lundvall and Johnson underline that 'the firms of the learning economy are to a large extent "learning organisations"' (Lundvall and Johnson, 1994, p. 26). A dynamic and flexible 'learning organisation' can be defined as one that promotes the learning of all its members and has the capacity of continuously transforming itself by rapidly adapting to changing environments by adopting and developing innovations (Pedler *et al.*, 1991; Weinstein, 1992). In order to meet the challenges of a global economy it is important that such learning organisations are embedded in broader social structures (Granovetter, 1985), either in the form of strong involvement of workers at the intra-firm level, of horizontal cooperation in networks at the inter-firm level, and of bottom-up, interactive based innovation systems at the regional level and beyond (Asheim and Isaksen, 1997; Ennals and Gustavsen, 1998).

Learning as a Localised Process

Lundvall and Johnson use the concept of a 'learning economy' when referring to the ICT (information, computing and telecommunication) -related techno-economic paradigm. They emphasise that 'it is through the combination of widespread ICT-technologies, flexible specialisation and innovation as a crucial means of competition in the new techno-economic paradigm, that the learning economy gets firmly established' (Lundvall and Johnson, 1994, p. 26). These perspectives of the 'learning economy' are based on the view that knowledge is the most fundamental resource in a modern capitalist economy, and learning the most important process (Lundvall, 1992), thus making the learning capacity of an economy of strategic importance to its innovativeness and competitiveness.

One of the consequences of knowledge-intensive modern economies is that 'the production and use of knowledge is at the core of value-added activities, and innovation is at the core of firms' and nations' strategies for growth' (Archibugi and Michie, 1995, p. 1). Thus, in a 'learning economy',

... technical and organisational change have become increasingly endogenous. Learning processes have been institutionalised and feed-back loops for knowledge accumulation have been built in so that the economy as a whole ... is 'learning by doing' and 'learning by using' (Lundvall and Johnson, 1994, p. 26).

The perspective of 'learning economies' and modern innovation theory emphasises that learning is a localised, and not a placeless, process (Lundvall and Johnson, 1994; Storper, 1995a). This view is supported by Porter, who argues that 'competitive advantage is created and sustained through a highly localised process. Differences in national economic structures, values, cultures, institutions, and histories contribute profoundly to competitive success' (Porter, 1990, p. 19). Accordingly, Porter argues that 'the building of a "home base" within a nation, or within a region of a nation, represents the organiza-tional foundation for global competitive advantage' (Lazonick, 1993, p. 2). In contrast to this, Reich (1991) argues that 'the globalization of industrial competition has led to a global fragmentation of industry, thus making national industries and the national enterprises within them less and less important entities in attaining and sustaining global competitive advantage' (Lazonick, 1993, p. 2). According to Reich, 'the work of nations' is the result of activities that take place in the national territory, and not of nationally-based companies (Reich, 1991; Storper, 1995b).

Storper considers the views of Reich and Porter, respectively, as:

> ... two widely differing analyses of how policy could implement a learning economy. One (Reich) can be characterised as 'global economy + the generic public good of labor' and the other (Porter) as 'national economy + the specific public good of technology'. The first leads to broadly-based competitiveness policies; the second leads to more focused technology policies (Storper, 1995b, pp. 291-292).

Reich points at 'embodied knowledge' as the most important factor in securing a nation's future prosperity, i.e. knowledge embodied in production equipment (hardware), which can be operated on the basis of universal codified knowledge with a general, global accessibility (software). Reich stresses especially the role of the quality of the work force, arguing that human capital investments are the most efficient public policy for attracting high-wage and high-value-added activities, demanding high-skill labour, to advanced nations from the 'global webs' of TNCs (Lazonick, 1993; Reich, 1991; Storper, 1995b). Porter focuses on the importance of 'disembodied knowledge' in promoting innovativeness and competitiveness, i.e. knowledge and know-how which are not embodied in machinery.

Reich's analysis, in contrast to Porter's, partly misses the importance of territory (i.e. location and agglomeration) and non-economic factors (i.e. institutions, social structures, traditions etc.) for the performance of an economy. Moreover, Reich's policy does not look especially promising from the point of view of keeping advanced, high-cost, welfare economies on a high-wage/high innovation development path, when the contemporary developments in the global economy such as the rapidly increasing competitive advantage of, for example, the Indian software industry, based on low paid engineers, and the heavy investments in higher education (especially applied science such as information technology) in countries like South Korea and Taiwan are taken into consideration (cf. Fromhold-Eisebith, Chapter 9, *this volume*).

The major impact of Porter's book, The *Competitive Advantage of Nations* (1990), is represented by a change in the understanding of the strategic factors which promote innovation and economic growth. Porter's main argument is that these factors are a product of localised learning processes, and that the importance of clusters is that they represent the material basis for an innovation based economy, which represents 'the key to the future prosperity of a nation' (Lazonick, 1993, p. 2).

Regional Agglomerations

A strong case is made today that regional agglomeration is growing in importance as a mode of economic coordination (Asheim and Isaksen, 1997; Cooke, 1994). The main argument for this is that regional agglomeration provides the best context for an innovation-based learning economy. In general, 'geographical distance, accessibility, agglomeration and the presence of externalities provide a powerful influence on knowledge flows, learning and innovation and this interaction is often played out within a regional arena' (Howells, 1996, p. 18).

This combines with a growth in the use of information and communication technologies, decentralisation of industry policies, and the emergence in accomplished regional economies of associative governance systems (Cooke and Morgan, 1998), where private bodies take some responsibility for policy development and administration (e.g Chambers of Commerce; regional conferences; consortia and forum arrangements for discursive exchange at local and regional levels).

At the regional level innovative capacity can be promoted through identifying 'the economic logic by which milieu fosters innovation' (Storper, 1995c, p. 203). Generally, it is important to underline the need for 'enterprise support systems, such as technology centres or service centres, which can help keep networks of firms innovative' (Amin and Thrift, 1995, p. 12; cf. Crevoisier, Chapter 3, *this volume*). This points to the importance of *disembodied knowledge* in promoting innovativeness and competitiveness, i.e. positive externalities of the innovation process in the form of knowledge and know-how (not tied to the use of, e.g., a particular machinery), as well as to *untraded interdependencies*, i.e. 'a structured set of technological externalities which can be a collective asset of groups of firms/industries within countries/regions and which represent country- or region-specific "context conditions" of fundamental importance to the innovative process' (Dosi, 1988, p. 226).

Disembodied knowledge can be both tacit and codified. However, such knowledge is generally based on 'a high level of individual technical capacity, collective technical culture and a well-developed institutional framework ...

[which] ... are highly immobile in geographical terms' (de Castro and Jensen-Butler, 1993). Thus, codified knowledge can be a product of localised learning rather than a result of learning based on ubiquitous factors (cf. Maskell, Chapter 2, *this volume*). This implies that the adaptability of this localised form of codified knowledge is dependent upon, and limited by, contextual, tacit knowledge, and, thus, represents important 'context conditions' of regional agglomerations with a potentially, favourable impact on their innovativeness and competitiveness (Asheim and Cooke, 1998).

This localised form of codified knowledge can provide an improved basis for 'learning by interacting' (e.g. user-producer relationships), which represents a more advanced form of learning than 'learning by doing' and 'learning by using'. 'Learning by interacting' cannot, in order to be fully exploited, only be based on tacit knowledge. Relying on the emphasis modern innovation theory places on interactive learning, it could be argued that the combination of *territorially embedded Marshallian agglomeration economies, the interplay of tacit and codified disembodied knowledge* and *untraded interdependencies* could constitute the basis for a new form of socially created comparative advantage for regions in the globalised economy (cf. Maskell, Chapter 2, *this volume*). Thus, it seems possible to argue that the most viable alternative for advanced countries is a policy of strong competition (Storper and Walker, 1989), i.e. a competition building on innovation and a 'differentiation' strategy. Such a strategy is based on localised learning in industrial clusters and territorial agglomerations. Policy instruments should be created aiming at a systematic promotion of localised learning processes in order to secure the innovativeness and competitive advantage of an economy (Freeman, 1995). This would represent a complement to the argument that 'ubiquitification', as an outcome of globalisation and codification processes, in general will undermine the competitiveness of firms in high-cost regions and nations (cf. Maskell, Chapter 2, *this volume*).

The Weakness of Regional Agglomerations: Lack of Innovative Capacity

Marshall was aware of the fact that agglomeration economies as such do not guarantee that product and process innovations will take place. Studies have shown that the 'industrial atmosphere' of industrial districts can support the adoption, adaptation and diffusion of innovations among SMEs (Asheim, 1994). The presence of trust can stimulate the introduction of new technology into industrial districts, since mutual trust seems to be crucial for the establishment of non-contractual inter-firm linkages.

The importance of territorial agglomerations in promoting innovations concerns largely incremental innovations. This is especially the case when regional economies are dominated by clusters of SMEs. Even if such incremental innovations individually have no major impact, their combined effect can be extremely important for product design and productivity growth in different branches, especially in relation to the overall economic performance of SMEs. Freeman underlines 'the tremendous importance of incremental innovation, learning by doing, by using and by interacting in the process of technical change and diffusion of innovations' (Freeman, 1993, pp. 9-10).

Thus, agglomeration economies can represent important basic conditions and stimulus to incremental innovations through informal 'learning-by-doing' and 'learning-by-using', primarily based on tacit knowledge (Asheim, 1994). As Bellandi suggests, such learning, based on practical knowledge (experience) of which specialised practice is a prerequisite, may have significant creative content, implying that the collective potential innovative capacity of small firms in industrial districts is not always inferior to that of large, research-based companies (Bellandi, 1994). The fact remains, however, that, in general, the results of decentralised industrial creativity are incremental at the level of individual firms, even if 'their accumulation has possible major effects on economic performance' (Bellandi, 1994, p. 76) at the regional scale.

In an increasingly globalised world economy it is rather doubtful whether incremental innovations will be sufficient to secure the necessary competitiveness of territorially agglomerated SMEs. Crevoisier argues that the reliance on incremental innovations 'would mean that these areas will very quickly exhaust the technical paradigm on which they are founded' (Crevoisier, 1994, p. 259), and Bellandi sees 'the assessment of the endogenous innovation capacities of the industrial districts . . . [as] . . . a key issue' (Bellandi, 1994, p. 73). More specifically, this means the capability to break path dependence and change technological trajectory through radical innovations, so as to avoid falling into 'lock-in situations' as a result of internal 'weakness of strong ties' (Granovetter, 1973) or of external 'weak competition' from low cost producers (Glasmeier, 1994). Crevoisier emphasises the importance of understanding how industrial districts 'react to or generate radical innovations. Without making this point clear, it is not possible to make any prediction about the reproduction and the duration of such systems' (Crevoisier, 1994, p. 259).

The strict dichotomy often applied between codified and tacit knowledge can be quite misleading both from a theoretical as well as from a policy point of view. This is especially the case if localised learning is said to be based primarily on tacit knowledge. A claim for the superiority of tacit knowledge on such a ground could lead to a fetishisation of the potentials of local production systems, and a failure to discover the problems such systems could face due to their lack of strategic, goal-oriented actions and strategies, which always have to be based on codified knowledge (e.g. formal R&D) (Amin and Cohendet, 1997). The category of disembodied, codified knowledge represents a concept which grasps the important basis for endogenous regional development: it refers to localised learning which is built on a strategic use of both codified, (possibly) R&D-based knowledge and tacit knowledge.

The 'learning economy' perspective focuses primarily on 'catching up' learning (i.e. learning by doing and by using) based on incremental innovations, and not on radical innovations (Lundvall, 1992; Lundvall and Johnson, 1994). What is needed to stay competitive in a global economy is the creation of new knowledge through searching, exploring and experimentation involving creativity as well as more systematic R&D in the development of new products. This goes for both semi-customised, design-intensive consumer goods and research-intensive capital goods.

One way of solving the problem of improving the innovative capacity of the small-firm sector of industrial agglomerations – to avoid these firms remaining as firms with a low level of internal resources and competence – is to rely on collective capacity-building by setting up centres for real services which could systematically assist firms in industrial agglomerations so that they are able to keep pace with the latest technological development. This could be done either through a networking strategy between firms and public and private agencies, or through public intervention. However, for SMEs to carry out (especially radical) innovations there is often a need to supplement the informal, tacit and localised form of codified knowledge with R&D competence and more systematically accomplished basic research, typically taking place within universities and research institutes. In the long run, most firms cannot rely only on localised learning, but must also have access to more universal, codified knowledge of, for example, national innovation systems. The strength of the traditional, place-specific and often informal competence and tacit knowledge must be integrated with more generally available and possibly R&D-based, codified knowledge. According to Varaldo and Ferrucci, 'long-term strategic relationships, R&D investments, engineering skills, new technical languages and new organizational and inter-organizational models are needed for supporting these innovative strategies in firms in industrial districts' (Varaldo and Ferrucci, 1996, p. 32).

In order to keep their position in the global market, firms must focus on developing their core competencies through transforming themselves into learning organisations as well as engaging in 'global clusters'. But internal restructuring alone cannot sustain the competitiveness of firms in the long run. Firms are very much dependent on favourable economic environments. In general, close cooperation with suppliers, subcontractors, customers and support institutions in the region will enhance the process of interactive learning and create an innovative milieu favourable to innovation and constant improvement. This influences the performance of the firms and strengthens the competitiveness of the regional agglomeration.

Innovation Networks

In the promotion of a regionalisation approach emphasising strong competition, the formation of innovative networks plays an important role. Strong competition means high innovation, high skills and high incomes rather than price competition and low wages. An 'innovation network' is a means for organising innovative capabilities and activities. As communication, cooperation and coordination amongst actors are conceived of as preconditions for the generation and dissemination of new commercial products and services, the perspective on innovation policies is changed from one which stresses traditional physical and functional infrastructure to one which focus on knowledge infrastructures in order to increase innovation capability and foster interactions between firms and other. Cooperation through networks helps reduce uncertainties and helps smaller firms incapable of 'worldwide sourcing' to engage in collaborative problem solving (Körfer and Latniak, 1994).

The regional dimension is crucial for innovation networks for several reasons:

- the capacity for developing human capital is facilitated by interactions between nearby firms, schools, colleges and vocational training mediators;
- networks of both formal and, mainly, informal contacts between network members are made possible through the ease of casual or planned meetings, information exchanges and customer-supplier relationships;
- synergies, or an innovative 'surplus' that can arise from the shared cultural, psychological or political perspectives arising from occupancy of a shared economic space or region; and

- the frequent existence of legitimate, strategic administration powers in fields such as education, innovation and enterprise support.

Extensive studies conducted by Maillat (1991) and Perrin (1988) conclude that there are two main types of innovation network:

- *The endogenous innovative network* is based upon a pre-existing regionally or locally delineated cluster of small and medium enterprises. They will have had a lengthy tradition of interacting and learning from one another, successfully competing on the basis of, as needed, co-operative innovation practices. Examples of such endogenous innovative networks are to be found in southern Germany (e.g. Baden-Württemberg) and the Third Italy (e.g. Tuscany or Emilia-Romagna).
- *The exogenous innovative network* takes the form of technopoles or science parks. They tend to emerge in two kinds of circumstances: (a) when large firms fragment their production structure and locate R&D activities in functionally specialised sones where synergies are expected to arise from co-location (as in Sophia Antipolis or Lille in France), or (b) by planned innovative milieus established to promote collaboration between universities and SMEs (as in science parks in the UK and USA).

Falling between these two approaches are innovative networks which develop in or near already existing metropolitan areas and combine characteristics of both the endogenous and exogenous type. Large and smaller scale firms establish network relationships with universities, other firms, research institutes, and government agencies. They benefit from policy networks supportive of innovation processes of network members.

Through networking the aim is to create 'strategic advantages over competitors outside the network' (Lipparini and Lorenzoni, 1994, p. 18). To achieve this it is important that the networks are organised so as to support the 'strength of weak ties' (Granovetter, 1973). Grabher argues that 'loose coupling within networks affords for favorable conditions for interactive learning and innovation. Networks open access to various sources of information and thus offer a considerably broader learning interface than is the case with hierarchical firms' (Grabher, 1993, p. 10).

It has been noted that:

... subcontracting relationships ... have changed substantially: they are no longer confined to the goal of cost savings only, but increasingly include aspects of product quality and technology development and improvement. This implies more selective and fewer but stronger relationships between firms since they cover not just production but also quality control, joint research and development as well as information exchange on and coordination of future planning (Tödtling, 1995, p. 14).

The spatial proximity between interacting firms is an important enabling factor in stimulating inter-firm 'learning networks' involving long-term commitment. The reorganisation of networking between firms can be described as a change from a domination of vertical relations between principal firms and their subcontractors to horizontal relations between principal firms and suppliers. Patchell refers to this as a transformation from production systems to learning systems, which implies a transition from 'a conventional understanding of production systems as fixed flows of goods and services to dynamic systems based on learning' (Patchell, 1993, p. 797).

Of additional importance is the formation of innovation networks between actors in territorial agglomerations and external economic actors, giving priority to horizontal inter-firm technological cooperation to ensure the adoption and diffusion of radical innovations. According to Camagni, innovation networks with external and specialised milieus may provide local firms with 'the complementary assets they need to proceed in the economic and technological race' (Camagni, 1991, p. 4).

With this perspective on networks, 'new competition . . . is among alliances of firms, not individual firms' (Rosenfeld, 1996, p. 247; Gomes-Casseres, 1994). Competitive advantage is achieved within regional networks through inter-firm cooperation and exploited externally through collaboration and/or competition with firms of the 'outside' world. '[T]o fight foreign rivals requires a suspension of rivalry in order to build value-creating industrial and technological communities. Unless social organizations are put in place that can engage in innovation, heightened domestic rivalry will lead to decline' (Lazonick, 1993, p. 8).

Towards Regional Innovation Systems

Building upon and extending key features of the localised interactive innovation networks with respect to partnership, flexibility, innovation support, linkage, variety, foresight, hard and soft infrastructures, the notion of a regional innovation system involves a strategic institutionalisation of innovation between the private and public sectors in a systemic way, constituting an institutional infrastructure as a 'superstructure' to the production structure of a region. Taking each element of the term Regional Innovation System (RIS) in turn, the concept *region* recognises the existence of an important level of industry governance between the national and the local. To varying degrees, regional governance is expressed in both private representative organisations such as branches of industry associations and chambers of commerce, and public organisations such as regional ministries with devolved powers concerning enterprise and innovation support, particularly for SMEs. Furthermore, there are few regions that do not possess increasingly important universities or polytechnics that can look outward to industry either for research commissions or as incubators for innovative start-up firms.

'Innovation' refers to the process of commercialising new knowledge, possibly though not necessarily emanating from universities, with respect to product, process or organisational innovation. As we have seen, this is now better understood as a complex process involving users, producers and various intermediary organisations learning from each other regarding demand and supply capabilities and exchanging both tacit and codified knowledge. Indeed, innovation can be characterised as a knowledge-transfer and realisation process involving actors whether internal or external to the specific firm operating as a project-based team or project-network. The *systemic* dimension of the 'RIS' derives in part from this team-like character associated with innovation in networks. While, as Lundvall (1992) puts it, an innovation system is a set of relationships between entities or nodal points involved in innovation, it is really much more than this. Such relationships, to be systemic, must involve some degree of inter-dependence; not all relationships may be equally strong all of the time, but some may be. Likewise, not all such systemic relations need be regional, but many are, and as out-sourcing grows more are likely to become so.

It is largely as a consequence of the increased externalisation of production and services (including research) into often regionalised 'clusters' that the management of innovation has become organisationally more complex. Whereas Lundvall and Johnson (1994) have written incisively about the 'learning economy' being rooted in 'learning organisations', the real implication for learning economies is that they must aspire to the control and support structures internalised in learning organisations, in the externalised environment of the 'learning region' (Florida, 1995; Storper, 1995c; Asheim, 1996). But we must move even a step further than this, for innovation involves more than learning, it requires creativity and even an aspect of 'tutoring' when it comes to the important task of identifying unrecognised needs for support especially by SMEs (Cooke and Morgan, 1998). It is when a regional economy has evolved to this point that it warrants the designation of a regional innovation system. The externalised control and support of the necessary flows of authority, responsibility, reputation, trust, information and finance through efficiently functioning networks, alert and alerted to the nuances of innovative actions on a global scale, is beyond the scope of most single firms. It requires a systemic, networked approach which brings together regional governance mechanisms, universities, research institutes, technology-transfer and training agencies, consultants and other firms acting in concert on innovation matters.

Some regional systems are better equipped to do this than others as has recently been argued (Cooke *et al.*, 1998), as systemic relationships can also exemplify 'lock-in' and 'the weakness of strong ties' (Grabher, 1993; Granovetter, 1985). In ideal-typical terms Table 6.1 summarises aspects of this difference.

This points to the importance of the *knowledge infrastructures* of regions and countries. According to Smith (1997), 'any analysis of the technological performance of a country or region should therefore have the infrastructure clearly in focus' (Smith, 1997, p. 94). Knowledge infrastructures are comprised of a variety of organisations such as universities, other R&D institutions, training systems, and firms, 'whose role is the production, maintenance, distribution, management, and protection of knowledge' (Smith, 1997, pp. 94-95).

Such knowledge infrastructures are of strategic economic importance concerning the promotion of innovation and economic growth, since all industrial production is based on knowledge, which can be either formal, codified (scientific or engineering knowledge) or informal, tacit (embodied in skilled personal routines or technical practice) (Smith, 1997). In this chapter we have, in particular, focused on the importance of what could be called 'soft' knowledge infrastructures, i.e. infrastructures producing knowledge according to an interactive, bottom-up model, for regional economic performance.

'Soft' knowledge infrastructures can be associated with Lundvall's 'broad' definition of innovation systems, which 'includes all parts and aspects of the economic structure and the institutional set-up affecting learning as well as searching and exploring' (Lundvall, 1992, p. 12). This broad definition

Table 6.1   Characteristics of regions with strong and weak systemic innovation potential

Strong regional innovation potential	Weak regional innovation potential
Infrastructural level	
Autonomous spending and taxation	Decentralised spending and taxation
Regionalised private finance	National private finance
Strategic infrastructure competence	Few infrastructure competences
Embedded universities/R&D labs	Disembedded universities/R&D labs
Organisational level: Firms	
Workplace cooperation	Workplace antagonism
Externalisation	Internalisation
Innovation	Adaptation
Organisational level: Policy	
Inclusive	Exclusive
Monitoring	Reacting
Consultation	Authorisation
Institutional level	
Consensus	Dissensus
Associative	Individualistic
Learning disposition	Introspective

incorporates the elements of an interactive innovation model, in contrast to the narrow definition, which is based on the linear model of innovation.

Thus, it is important, analytically as well as politically, to distinguish between different types of regional innovation systems, as a parallell to the two main types of innovation networks presented earlier. On the one hand, we find innovation systems that could be called *regionalised national innovation systems*, i.e. parts of the production structure and the institutional infrastructure located in a region, but functionally integrated in, or equivalent to, national (or international) innovation systems, which are more or less based on a top-down, linear model of innovation. On the other hand, we can identify *territorially integrated innovation systems* constituted by the parts of the production structure and institutional set-up that are territorially integrated, and built up by a bottom-up, interactive innovation model.

Examples of a *regionalised national innovation system* could be the R&D laboratories of large firms, governmental research institutes or 'science parks', often located in the proximity of technical universities and based on the thinking of the linear model of innovation, and with rather limited linkages to local industry (Asheim, 1995; Henry *et al.*, 1995). This all implies a lack of territorial embeddedness and leads to questions about their capability for promoting innovativeness and competitiveness on a broad scale in local industries (especially the SMEs) in particular regions, as a prerequisite for endogenous regional development. However, there is normally better networking between R&D institutions, firms and the local state in regionalised national innovation systems than in national ones, e.g. the regional innovation system of Baden-Württemberg (Cooke and Morgan, 1994).

The best examples of *territorially integrated regional innovation systems* are networking SMEs in industrial districts, which build their competitive advantage on localised learning processes. Thus, in Emilia-Romagna, for example, the innovation system could be said to be territorially embedded within that particular region. The rationale for such territorially-embedded systems is to provide a bottom-up, network-based support (e.g. through technology centres, innovation networks or centres for 'real services') for the 'adaptive technological and organizational learning in territorial context' (Storper and Scott, 1995, p. 513). To be able to talk about a territorially-integrated regional innovation system the national, functionally integrated, techno-economic and political-institutional structures must be 'contextualised' through interaction with the territorially embedded, socio-cultural and socio-economic structures (Asheim, 1995).

However, different industrial sectors, in terms of size and forms of organisation have different requirements of innovation systems and innovation policy. There are obviously differences in demand between *locally controlled SMEs* (including their subcontractors/suppliers), *large locally-controlled firms* (including their supply chains), *subcontractors/suppliers for firms outside the region*, and *branch plants*. While the first category of firms primarily need the support of an interactive, regionally embedded innovation system, the last three categories of firms largely demand their services from national and international, sectoral innovation systems. The second category of large, locally-controlled firms can, of course, also make use of a regionalised national innovation system. In addition, depending on the sectors of the firms, the services of a regionally embedded innovation system can be used. Even in an

R&D-dominated industry such as the petro-industrial complex, the mechanical firms of the offshore-industry can benefit from the broader view of interactive learning as central to innovative activies, and especially by exploiting the increased impact of a territorial agglomeration of a local production system on the firms' international competitive advantage (Asheim and Isaksen, 1997).

Examples of Interactive-Model Innovation Networks

The Italian *industrial districts* exemplify a strong local support of technology and innovation. Enterprise support of a more proactive kind has been induced recently by agreement amongst small firms, the regional government and the intermediary agencies. The period of conscious intervention accelerated during the 1990s when the ERVET (Ente Regionale per la Valorizzazione Economia del Territorio – the regional development agency) system of enterprise support was put into place to assist the networked firms to deal with competition based on advanced technologies, on the one hand, and cheap overseas labour, on the other. Amongst these new initiatives are the following which must be considered of importance:

• *Technology Centres*: where smaller firms may access technology services; R&D information; professional training; high-quality communication services; management advice; marketing support.
• *Advanced Telecommunications*: optical fibre networks; ISDN; broadband services; specialist business services and shared facilities/costs; and video conferencing.
• *Access to Technology Foresight and Monitoring*: through networked information and services provision, including gateway services to international markets.

In other words, a large-scale investment in innovation networks must provide both hard and soft infrastructure in terms of wiring and human-centred support services to make it an attractive location for innovative firms.

The *'network' approach* is more typical of Germany, Austria and the Nordic Countries and distinct from the endogenous, Italian industrial district approach to technology and innovation support as well as the linear-model, exogenous approach, of which science parks and technopoles are most representative. In general science parks tend to have weak local cooperative environments (Henry *et al.*, 1995), which result in a failure to develop inter-firm networking and interactive learning in the parks, while technopoles are characterised by a limited degree of innovative interaction between firms in the poles, and by vertical subcontracting relationships with external firms (Asheim and Cooke, 1998).

In this section, a number of cases from these countries will be outlined, pointing to the planned, interactive enterprise-support approach relying on close university-industry cooperation. First we will examine a larger-scale interactive approach in Germany, and a smaller one in Austria. Then we move on to the experiences of some Nordic countries, notably Sweden and Finland, where successful innovative environments are of special interest. They tend to involve a Technical University, new start-up firms in a neighbouring Technology Park and the close presence of a large firm that, through its R&D laboratory, links to the University scientists and provides a market for the products and services established by academics or their students. In this way vertical and horizontal linkages among larger and smaller innovation actors ensue. Examples of this can be found in Linköping (Sweden) and Oulu (Finland).

North Rhine-Westphalia, Germany

North Rhine-Westphalia has sought to develop innovation support policies since 1972. The NRW Technical Board established then an Innovation Centre (ZENIT), based at Mülheim, and others for the Ruhr and for the whole of the North Rhine-Westphalia Land in the later 1970s and 1980s (Kilper and Fürst, 1995).

In 1987, the land had set up the Zukunftsinitiative Montanregionen (ZIM, Initiative for the future of the coal and steel regions). Local actors were brought together and asked to decide, by consensus, which projects should be proposed to the land for funding. Rather than a programme being set up by land or local government, a broad range of actors in each locality, e.g. Chambers of Commerce and Trade, banks, local politicians, were brought together to form 'regional conferences'. These bodies had, and still have, no official or decision-making powers, but rather work alongside existing tiers of local government [Regierungsbezirke, Kreise/cities, Gemeinde which may be compared to counties, (metropolitan) districts, communities]. Under ZIM, the regional conferences decided, by consensus, on projects for their localities, which were then proposed to the land for funding. Action areas were: innovation and technology; training for the future; infrastructural modernisation; and improvements to energy and the environment. The land funded 300 of 1200 projects. In 1989, 'Zukunftsinitiative für die Regionen Nordrhein-Westfalens' (ZIN 1, Initiative for the future of NRW's regions) was set up for all the land's 15 regions. Regional fora were set up, involving a broad range of actors within localities e.g. local government, Chambers of Commerce, industry, and banks.

Under ZIN 1, greater emphasis was laid on strategy than under ZIM, but the same action areas applied. The land gave DM 1.1 billion to 330 projects out of 2,000 proposed. ZIN 2 followed in 1990, when the land decided that each region should compile a 'Regionale Entwicklungskonzept' (REK, Regional development programme) based on an empirical analysis of its economic situation, and setting out a strategy for future development.

It has recently been proposed that the REKs should move towards forming the land's regional policies. This is a significant development, because it gives considerably more strategic power to non-elected regional conferences or fora. Such conferences will propose projects to the land government which, if approved, will commit the regional administrative authorities within the land to carrying them out. The land controls this process through allocating resources earmarked for the proposed projects to the regional administrations for them to carry out even though they have not actually initiated them.

In November 1991, the 'Initiative Bergbautechnik' (Mining technology initiative) was set up by the Ministry of Economics and the Ministry for Work, Health and Social Affairs in order to facilitate the implementation of structural changes in the Emscher-Lippe region, the eastern Ruhrgebiet and Aachen-Heinsberg's coal area. The private sector participates in the programme (both *Mittelstand* and large firms, as well as the Chambers of Commerce and industry). The 1992-95 programme was funded by the European Community (EC).

A number of technology centres have been set up under the NRW Technology Programme. These have the aim of bringing together innovative and technological activities and firms in the hope of creating synergies, which will then have effects throughout the region. The centres are part of a process of restructuring and modernising the land economy, helping to create highly-qualified, technology-oriented jobs. Firms are offered a range of services, including cheap office, R&D and production space; reception and telephone facilities; cafeteria, business advice and conference rooms. Networking and exchange of ideas and information are seen as crucial to the success of the centres. The state is perceived as having an important facilitative role in this process. However, initiatives to set up technology centres usually come from the local - grassroots - level and generally take the form of public private partnerships. At present there are 31 Technology Centres in NRW and a further 12 are projected for establishment in the near future.

Learning and improvement of network mechanisms have occurred during the lengthy period during which the North Rhine-Westphalia programme has been in operation. In brief, these can be summarised as follows:

- To assist in keeping new companies in the region it is important to link any centre (higher education institute, science park) responsible for their establishment with a technology park where they can, for comparable rents, turn research into product.
- The network should not focus on a few technological areas. This is because of the accepted difficulty in 'picking winners'. In any case, networks of the kind under discussion tend not to be highly specialised. A 'technology mix' concept in which there is variety in the fields of expertise is to be favoured.
- Professional management of the network is absolutely essential. Personnel with experience of running a company, running the technology development area of a large company, or engaging in consultancy in technology fields of direct relevance to the network are optimal.
- There needs to be a regional technology development strategy or Regional Technology Plan for the promotion of an innovation infrastructure. Integration of local actors and shareholders of the network organisation is crucial.
- A survey of existing infrastructure and likely customer demand for innovation services is a prerequisite for establishing a facility likely to be in demand by firms in the innovation market place.

Moreover, change in the development of policies is something requiring acceptance by network managers. The example of PlaNet Ruhr is instructive in this respect. Here, the basic idea was to integrate scientific organisations, consultants, chambers of commerce and other relevant agencies in a regional network of organisations. The intention was that it should disseminate information and know-how assisting business restructuring and, in the process, introduce one or more of ten possible new processes in production. It proved difficult to integrate the network; people involved changed, some network members could not devote sufficient time, and training offers often did not fit the needs of the firms. The network was in danger of disintegrating. But instead of walking away, the members reorganised the network by separating different functions. The consultants worked in their sphere, the trainers in theirs and they were able to speak in a more focused way to their clients and the policy network behind the initiative. The network and the initiative still function after a number of years and co-ordination is now more appropriate to firm requirements.

Technical University of Graz, Austria

In 1993 a project was established to provide active technology transfer from the university to both start-up and established regional enterprises. The partners were the university and the city council. The key aims were to identify some 70 firms suitable for this, to visit them and market the relevant services of the university to them and to stimulate cooperation between them and scientists. This involved computerised identification of know-how, a company audit, and problem identification and solution with consequent after-care. During the 70 meetings some 200 concrete requests for knowledge transfer were identified and solutions took the following form:

- providing access to research thesis findings of relevance;
- informal consulting by a university consultant-pool;
- use of university technology services;
- contracts for small scale research by university research assistants;
- job offers from firms for students and alumni.

In addition, Graz has five technology parks for new business start-ups. New start-ups are encouraged from amongst graduating Ph.D. students. Each June some one hundred are invited to an Open Day on 'Setting up a new technology-based firm'. They are addressed by previous students who have successfully established businesses and, on average, three new start-ups are established per year. They receive a low-rent unit on the Technology Park at the university provided their new business is not directly competing with an existing firm. In this way new spin-offs are protected from predatory competition from the outset. They are encouraged to interact, learn and even cooperate with complementary firms in the informal approach the Austrians call 'coffee-break knowledge transfer'. Over one hundred new start-ups are now in existence at Graz (Cooke, 1996).

The University of Oulu, Finland

The Oulu case is interesting enough to have been profiled in the *Financial* and *Business Week* because of its location close to the Arctic Circle and because it represents the largest concentration of high-tech firms in Finland (300 in 1996, of which 100 start-ups date from 1985 at the earliest (Jussila and Segerståhl, 1997). The University of Oulu is largely responsible, having set up the Technical Research Centre in 1974, the Oulu technology park in 1982, and the Medipolis medical science park in 1990.

However the technology/medical science parks do not only house small start-up firms. Nokia brought its first operations to the technology park in the 1970s and now employs 5,000 in R&D and the production of base stations for mobile telephony. Some of Nokia's sub-suppliers in printed circuit boards, base station technology and electronics systems have followed. But this does not edge out the start-ups, rather it provides many of them with an immediate market through local sub-contracting opportunities. Hence a virtuous circle of interaction now exists between large telecommunications firms, smaller start-ups and the Technical Research Centre, with systemic knowledge-transfer amongst them.

The Medipolis has some 50 firms; most have originated with Ph.D. graduates establishing businesses researching and producing advanced medical equipment and products. Because medical technology is a global business, with the USA and other European health systems being major purchasers, familiar with buying from and collaborating with innovative spin-offs, there are no large firms on the Medipolis. In 1996 employment on the Technology Park (without Nokia) was 1,200 and at the Medipolis, 300 (Jussila and Segerståhl, 1997).

Linköping University, Sweden

This university was founded in 1972, its largest faculty being the Institute of Technology, covering 40 fields in eight departments. The university has established and encouraged a tradition of technology-transfer by staff. In 1984 a group of entrepreneurs and start-up owner-managers joined forces with the university to establish the Foundation for Small Business Development (SMIL), not least because some 40 spin-off firms had been established in the early 1980s, mostly from the university but some from Ericsson and Saab, located near the university. SMIL offers membership to small technology-based firms and enterprise support groups. It now has 150 members.

Membership costs £150 annually. The first SMIL activities involved building a network of technology-based entrepreneurs, advising them, promoting exchanges, assisting with management resources and providing marketing support. SMIL has a separate secretariat at the university.

Working closely with SMIL is the university's Centre for Innovation and Entrepreneurship (CIE), the task of which is to stimulate growth amongst the SMIL-members through new business development programmes, problem-solving groups of owner-managers, management training and club/networking activities. This approach, centred on the local Mjardevi Science Park, has enabled academic-based firms to, in one case, reach the 800-employee mark, with three others employing over 100, and the rest of a total of approximately 100 surviving start-ups employing between one and five persons. In total, some 1,500 jobs in advanced, mainly information technology companies can be traced back to the activities of the Institute of Technology, SMIL and CIE at Linköping University. Again, as with Oulu, the co-location of university, large technology-intensive firms like Ericsson, Saab and some suppliers, and the innovative start-up firms constitutes a systemic innovation arrangement in which knowledge-transfer moves among the three kinds of partner, and small firms receive some security as a base for market growth from having, on the one hand, a customer market locally and, on the other, sources of knowledge and technology-transfer plus management advice close by (Jones-Evans and Klofsten, 1997).

Local Networking: The Case of Aarhus, Denmark

The Aarhus region, bordered by the towns of Randers, Silkeborg and Skanderborg, with 600,000 inhabitants, is one of the most important economic areas of Denmark. Together with the food technology and environmental science sectors which dominate the region, there are also numerous biotechnology companies, energy technology, electronics and software firms. Food processing is the largest industrial sector and accounts for more than a third of all employees in the private sector. More than 80% of employees in the private sector work in establishments with less than 50 employees. The remaining 20% generally work in medium-sized companies. There are only a few companies with more than 1000 employees.

In 1991, the local council of Aarhus launched an industrial development initiative known as 'Plan 2001' aimed at generating 20,000 new jobs through a structure of public/private dialogue and private/private interaction. The key elements of Plan 2001 include the introduction of the following:

- Growth groups;
- innovation contracts;
- business advisory agency;
- venture capital investment company;
- agri-food forum;
- international investment location initiative;
- establishment of 'Knowledge Centres'.

Although in principle the objective of Plan 2001 encompassed all firms, it sought to get initiatives up and running as quickly as possible in order to create early successes which would serve as good examples and thereby create snowball effects through the force of example. The basis for exerting influence consists of the promotion of 'organic' networks (Grabher, 1993) involving the groups, organisations and also companies who themselves have an ongoing contact with many companies, and who have a self-interest in strengthening their business clients. As a result they are in a position to stimulate awareness of initiatives via an indirect information exchange. Other candidates involved in this dissemination process are the larger companies with numerous subcontractors, wholesalers and retailers, public sector purchasers and also industrial and trade organisations, employers' associations and trade unions (Nielsen, 1994).

Technological Cooperation: The Case of Jæren, Norway

Although much smaller than industrial districts in the Third Italy, one of the best examples of an industrial district type development in Norway is Jæren, located south of Stavanger in the south-western part of Norway. Here an organisation called TESA (Technical Cooperation) was established by local industry in 1957, with the aim of supporting technological development among the member firms, which were small and medium-sized, export-oriented firms producing mainly farm-machinery. This has, among other things, resulted in the district today being the centre for industrial robot technology in Norway with a competence in industrial electronics/micro-electronics far above the general level in Norway. Furthermore, the use of industrial robots is much more widespread in this region than in the rest of Norway (i.e. approximately 1/3 of all industrial robots with only 3% of Norway's industrial employment).

In 1994 TESA had 13 member firms with more than 2800 persons employed and a turnover of 2.2 billion NOK. The TESA firms have overall a very high export share with an average of 63% (i.e. 1.4 billion NOK in 1992).

However, in some of the firms a far larger share is exported; three firms had an export share of more than 90% in 1992: Lærdal (medical equipment) 96%, ABB Flexible Automation (painting robots) 96%, and Kverneland (farm-machinery) 91% (increasing to 94% in 1993). According to the firms, without the inter-firm technological cooperation taking place within TESA, the development of this very strong competitive advantage would not have been possible.

As part of the work to promote the member firms' competitive advantage, TESA took active part in the establishment of JÆRTEK (Jæren's technology centre) in 1987. The aim of JÆRTEK is to offer training to prepare workers and pupils in technical schools for the advanced industrial work of tomorrow, and to secure the competence basis for a continued, rapid technological development. To achieve this, the first complete computer-integrated manufacturing (CIM) equipment in Norway was installed in JÆRTEK. Later the CIM concept has been diffused to several other member firms, among them Kverneland, which used the investment in CIM to combat the reduced demand for agricultural machinery in Europe through increased productivity and competitiveness. This strategy resulted in a strong increase in the turnover in 1994 and 1995, which made Kverneland the largest producer of ploughs in Europe.

The most well-known firm at Jæren is ABB Flexible Automation, which was called Trallfa Robot before it was bought by ABB in the late 1980s. At that time Trallfa Robot supplied around 50% of the European market for painting robots to the car industry. If ABB had applied their normal restructuring strategy, the robot production at Jæren should have been closed down, and moved to Västerås in Sweden, where the production of handling robots took place at a much larger scale. Instead, the production capacity at Jæren has been increased from 200 robots in the early 1990s to 600 in 1995, and to around 1000 in 1996. This means that ABB Flexible Automation today supplies 70% of the demand for painting robots in European car industry, and 30% in USA. The work force has been increased by 80 people from 1994 to 1996, reaching a total of 230 employed. With a turnover of around 290 NOK in 1995, it is the most profitable unit of ABB in Norway. In addition, the factory at Jæren has been upgraded to a so-called 'supplying unit' in the ABB corporation, and the production of handling robots has in part been transferred from Västerås to Jæren.

The success story of ABB Flexible Automation has partly to do with the informal, tacit knowledge and social qualifications of the work force (i.e. Marshall's 'industrial atmosphere' as a result of strong common values (i.e.

the Protestant work ethic) and close family ties in the communities, characterising the region). It also has partly to do with the localised form of codified knowledge constituted by the specific, disembodied knowledge about painting robots at the factory at Jæren, and the general, interactive learning-based innovations within robot technology in TESA, which represents region-specific 'untraded interdependencies'. These factors were recognised by ABB as being extremely important for the competitive advantage of ABB Flexible Automation (Asheim and Isaksen, 1995).

The close, horizontal inter-firm cooperation and interacting learning process, resulting in the development of core technologies (radical product and process innovations), existing in this district is rather unique in an international context. The technological cooperation was strongly dependent on the high level of internal resources and competence of the firms, and did not originally involve R&D institutions in the regional 'capital' of Stavanger. However, in later years, regional and national R&D institutions have gradually become greater involved in the R&D work (e.g. Rogaland Reseach in Stavanger, Chr. Michelsens Institute in Bergen, and The Technical University and SINTEF (Center for Industrial and Technical Research) in Trondheim).

Commentary

It is clearly important to make innovative small-firm growth and development a systematic part of the soft infrastructure available in Technology Centres and Parks. The example of Linköping University with its membership-based Foundation for Small Business Development (SMIL) and Centre for Innovation and Entrepreneurship to help stimulate business growth through business development training, problem-solving groups and club or networking opportunities is very instructive, as well as Jæren with the interactive learning-based technological cooperation in TESA, in collaboration with R&D institutions belonging to both the regional and national innovation system, and the educational technological centre of JÆRTEK. So are the triangular links between university, Nokia and start-ups at the Oulu technology park, the Danish 'growth-groups' idea at Aarhus, the monitoring of SME needs by the Technology Centres at Graz in Austria, and the proximity of Saab and Ericsson to the firms on the technology park at Linköping. Each of these successful examples of apparently secure interaction for furthering innovation by linking supply and demand for know-how market opportunities shows the importance of the network approach to innovative business growth and development.

The strength of the network approach is based on the interactive innovation model, which points at cooperation and localised learning as key factors in promoting innovativeness and competitiveness. The cases also demonstrate the advantages of territorial agglomerations in stimulating interactive learning between universities and private companies. And finally, the planned, systemic elements found in many of the described cases emphasise the potentials of a fully developed regional innovation system of supporting localised learning.

So, putting together the main lessons from our cases, the following list covers many of the most important ingredients:

- Partnership amongst large, private firms, government, universities, intermediary agencies, research institutes and small firms.
- Clear and transparent management which is flexible and open, not bureaucratic and hierarchical.
- Soft infrastructure of enterprise support for business development and management training for technology growth and innovation.
- Polycentric linkage to other key nodes in the innovation system locally and globally.
- Industrial variety, but not comprehensiveness, to reduce the dangers of monocentric dependence.
- Intelligence functions aimed at anticipating future needs and opportunities through technology foresight.
- Advanced telecommunications infrastructure to maximise economies of time and minimise costs of transactions.
- Technology Centres to supply expert services for technology transfer from knowledge-centres such as universities and research institutes to small and large business enterprises and public organisations.

Conclusions

It has been argued that as the model which best explains innovation processes amongst firms and scientific organisations has shifted from linear to interactive, so the model for promoting regional and local economic development based on the promotion of innovation has moved from a hierarchical to a more networked one. It was pointed out that early attempts to implant innovation activities in selected geographical spaces, by encouraging decentralisation of research laboratories and innovative firms to technopole environments, often produced rather disappointing results, failing to achieve stated objectives regarding exploitation of a projected 'synergetic surplus' for innovation.

One reason why Marshallian industrial districts have remained competitive is that they inherited a cooperative culture from the civil societies within which they developed (Putnam, 1993). Now, though, tougher competition means that their capacity for localised learning through incremental innovations based on tacit knowledge is not enough. They have to be more research- and knowledge-oriented. This means that the firms that comprise them must look beyond centres of 'real services' to universities and research institutes – maybe outside the district – to access necessary knowledge.

It is important that knowledge infrastructures and innovation systems reflect this in the way they are designed and operated. Concerning the design, the multi-level aspects of knowledge infrastructures and innovations systems have to be acknowledged, i.e. the geographical scale and scope of the structures and systems ranging from the local to the global. Different industries, in terms of branch, size and forms of organisation, have different requirements with respect to knowledge infrastructures and innovation systems. Locally-controlled, traditional SMEs may benefit most from a regionally embedded innovation system, based on an interactive innovation model, while high-tech SMEs and large firms may need access to linear model-based, national innovation systems or transnational (e.g. EU) sectoral innovation systems. Concerning the operation of innovation systems, organisational innovations in order to promote national and sectoral innovation systems to apply a more interactive approach, would be an important action to take to increase the accessibility of such systems to traditional SMEs. In this way, the individual and collective needs of SMEs in different sectors and/or regions could be better targeted.

A major problem perceived by policy-makers in recent years has been that of improving links between the knowledge infrastructures and firms in general, and especially making the infrastructure more responsive to the individual and collective needs of SMEs. This calls for the development of a more coordinated policy with respect to the delivery of services to SMEs across a range of sectors. However, according to Smith, 'it is clear that any integrated approach to the knowledge infrastructure will require organizational innovation within the public sector itself' (Smith, 1997, p. 104).

Policies that sought to promote interaction between different innovation actors that had good reasons to interact, such as universities or research institutes, small start-up firms and larger customer firms, as practised in Scandinavia, Germany and Austria produced more satisfactory results in relation to less ambitious goals. A point has now been reached where innovative policy-thinking has evolved towards a broadening of the network approach to encompass regional innovation systems. These may embody localised interactive networks but also include the wider business community and governance structure to maximise the financial and associational assets of regions for the promotion of innovation. For the moment it seems that there is no single model of the successful regional innovation system. But a reasonably high degree of regional economic and policy autonomy, a willingness to recognise the multi-level nature of innovation governance, an inclusive and consultative policy mentality and an associative culture attuned to the importance of innovation for growth and jobs, are important ingredients in the successful promotion of innovation for the future.

This leads us to conclude that for maximum efficiency technological innovation, as a socially interactive process, needs to be organised in ways that maximise economic externalities from geographical proximity. This means creating situations where hard (technological) and soft (human) infrastructures and networks enable interaction to take place. Leading successful examples of this network-based form of nurturing a culture of collective entrepreneurship combine knowledge-centres (such as universities), incubators (to nurture small, high technology, spinoff firms), and a collaboratively-minded larger firm or firms that will act as the market for new businesses in the difficult start-up and early growth phases of development. This is not merely a theoretical proposal; it exists in reality in some new technology districts such as those we described in, for example, the Nordic countries. These shine an interesting and important light upon the newly-rediscovered role of a socially interactive culture of cooperation in the quest for competitive advantage.

References

Amin, A. and Thrift, N. (1995), 'Territoriality in the Global Political Economy', *Nordisk Samhällsgeografisk Tidskrift*, no. 20, pp. 3-16.

Amin, A. and Cohendet, P. (1997), 'Learning and Adaptation in Decentralised Business Networks', Paper presented at European Management and Organisations in Transition (EMOT) final conference, Stresa, Italy, September 1997.

Archibugi, D. and Michie, J. (1995), 'Technology and Innovation: An Introduction', *Cambridge Journal of Economics*, vol. 19, pp. 1-4.

Asheim, B.T. (1994), 'Industrial Districts, Inter-firm Co-operation and Endogenous Technological Development: The Experience of Developed Countries', in *Technological Dynamism in Industrial Districts: An Alternative Approach to Industrialization in Developing Countries?* UNCTAD, United Nations, New York and Geneva, pp. 91-142.

Asheim, B.T. (1995), Regionale innovasjonssystem - en sosialt og territorielt forankret teknologipolitikk? *Nordisk Samhällsgeografisk Tidskrift*, no. 20, pp. 17-34.

Asheim, B.T. (1996), 'Industrial Districts as "Learning Regions": A Condition for Prosperity?', *European Planning Studies*, vol. 4, pp. 379-400.

Asheim, B.T. and Cooke, P. (1998), 'Localized Innovation Networks in a Global Economy: A Comparative Analysis of Endogenous and Exogenous Regional Development Approaches', in F. Engelstad, G. Brochmann, R. Kalleberg, A. Leira and L. Mjøset (eds), *Comparative Social Research*, vol. 17, *Regional Cultures*, JAI Press, Stamford, CT, pp. 199-240.

Asheim, B.T. and Isaksen, A. (1995), Spesialiserte produksjonsområder mellom globalisering og regionalisering, in D. Olberg (ed.), *Endringer i næringslivets organisering*. FAFO-report 183, Oslo, pp. 61-97.

Asheim, B.T. and Isaksen, A. (1997), 'Location, Agglomeration and Innovation: Towards Regional Innovation Systems in Norway?', *European Planning Studies*, vol. 5, pp. 299-330.

Bellandi, M. (1994), 'Decentralised Industrial Creativity in Dynamic Industrial Districts', In *Technological Dynamism in Industrial Districts: An Alternative Approach to Industrialization in Developing Countries?* UNCTAD, United Nations, New York and Geneva, pp. 73-87.

Braczyk, H., Cooke, P. and Heidenreich, M. (eds) (1998), *Regional Innovation Systems*, UCL Press, London.

Brusco, S. (1990), 'The Idea of the Industrial District: Its Genesis', in F. Pyke, G. Becattini and W. Sengenberger (eds), *Industrial Districts and Inter-Firm Co-operation in Italy*, International Institute for Labour Studies, Geneva, pp. 10-19.

Camagni, R. (1991), 'Introduction: From the Local "Milieu" to Innovation through Cooperation Networks, in R. Camagni (ed.), *Innovation Networks: Spatial Perspectives*, Belhaven Press, London, pp. 1-9.

Cooke, P. (1994), 'The Co-operative Advantage of Regions', Paper prepared for the Harold Innis Centenary Celebration Conference on 'Regions, Institutions, and Technology: Reorganizing Economic Geography in Canada and the Anglo-American World', University of Toronto, September 1994.

Cooke, P. (1996), *Networking for Competitive Advantage*, National Economic and Social Council, Dublin.

Cooke, P. and Morgan, K. (1994), 'Growth Regions under Duress: Renewal Strategies in Baden-Württemberg and Emilia-Romagna', in A. Amin and N.

Thrift (eds), *Globalization, Institutions, and Regional Development in Europe*, Oxford University Press, Oxford, pp. 91-117.

Cooke, P. and Morgan, K. (1998), *The Associational Economy: Firms, Regions and Innovation*, Oxford University Press, Oxford.

Cooke, P., Uranga, M. and Etxebarria, G. (1998), 'Regional Systems of Innovation: An Evolutionary Perspective', *Environment and Planning A*, vol. 30.

Crevoisier, O. (1994), 'Book Review (of G. Benko and A. Lipietz (eds) *Les Regions qui Gagnent)*', *European Planning Studies*, vol. 2, pp. 258-260.

De Castro, E. and Jensen-Butler, C. (1993), 'Flexibility, Routine Behaviour and the Neo-classical Model in the Analysis of Regional Growth', Department of Political Science, University of Aarhus (mimeo).

Dei Ottati, G. (1994), 'Cooperation and Competition in the Industrial District as an Organization Model', *European Planning Studies*, vol. 2, pp. 463-483.

Dosi, G. (1988), 'The Nature of the Innovative Process', in G. Dosi, C. Freeman, R. Nelson, G. Silverberg G and L. Soete (eds), *Technical Change and Economic Theory*, Pinter Publishers, London, pp. 221-238.

Edquist, C. (ed.) (1997), *Systems of Innovation: Technologies, Institutions and Organizations*, Pinter, London.

Ennals, R. and Gustavsen, B. (eds) (1998), *Work Organisation and Europe as a Development Coalition*. John Benjamin's Publishing Company, Amsterdam and Philadelphia (forthcoming).

Florida, R. (1995), 'The Industrial Transformation of the Great Lakes Region', in P. Cooke (ed.), *The Rise of the Rustbelt*, UCL Press, London, pp. 162-176.

Freeman, C. (1993), 'The Political Economy of the Long Wave', Paper presented at EAPE conference on 'The Economy of the Future: Ecology, Technology, Institutions', Barcelona, October, 1993.

Freeman, C. (1995), 'The "National System of Innovation" in Historical Perspective', *Cambridge Journal of Economics*, vol. 19, pp. 5-24.

Glasmeier, A. (1994), 'Flexible Districts, Flexible Regions? The Institutional and Cultural Limits to Districts in an Era of Globalization and Technological Paradigm Shifts', in A. Amin and N. Thrift (eds), *Globalization, Institutions, and Regional Development in Europe*, Oxford University Press, Oxford, pp. 118-146.

Gomes-Casseres, B. (1994), 'Group versus Group: How Alliance Networks Compete', *Harvard Business Review*, vol. 72 (4), pp. 62-74.

Grabher, G. (1993), 'Rediscovering the Social in the Economics of Interfirm Relations', in G. Grabher (ed.), *The Embedded Firm: On the Socioeconomics of Industrial Networks*, Routledge, London, pp. 1-31.

Granovetter, M. (1973), 'The Strength of Weak Ties', *American Journal of Sociology*, vol. 78, pp. 1360-1380.

Granovetter, M. (1985), 'Economic Action and Social Structure: The Problem of Embeddedness,' *American Journal of Sociology*, vol. 91, pp. 481-510.

Henry, N., Massey, D. and Wield, D. (1995), 'Along the Road: R&D, Society and Space', *Research Policy*, vol. 24, pp. 707-726.

Howells, J. (1996), 'Regional Systems of Innovation?', Paper presented at HCM conference on 'National Systems of Innovation or the Globalisation of Technology? Lessons for the Public and Business Sector', ISRDS-CNR, Rome, April 1996.

Jones-Evans, D. and Klofsten, M. (1997), 'Universities and Local Economic Development: The Case of Linköping,' *European Planning Studies*, vol. 5, pp. 77-93.

Jussila, H. and Segerståhl, B. (1997), 'Technology Centres as Business Environments in Small Cities', *European Planning Studies*, vol. 5, pp. 371-384.

Kilper, H. and Fürst, D. (1995), 'The Innovative Power of Regional Policy Networks: A Comparison of Two Approaches to Political Modernisation in North Rhine-Westphalia', *European Planning Studies*, vol. 3, pp. 287-304.

Körfer, H. and Latniak, E. (1994), 'Approaches to Technology Policy and Regional Milieux: Experiences of Programmes and Projects in North-Rhine-Westphalia', *European Planning Studies*, vol. 2, pp. 303-320.

Lazonick, W. (1993), 'Industry Cluster versus Global Webs: Organizational Capabilities in the American Economy', *Industrial and Corporate Change*, vol. 2, pp. 1-24.

Lipparini, A. and Lorenzoni, G. (1994), 'Strategic Sourcing and Organizational Boundaries Adjustment: A Process-Based Perspective', Paper presented at the workshop on 'The Changing Boundaries of the Firm', European Management and Organisations in Transition (EMOT), Como, Italy, October 1994.

Lundvall, B.-Å. (1992), 'Introduction', in B.-Å. Lundvall (ed.), *National Systems of Innovation*, Pinter, London, pp. 1-19.

Lundvall, B.-Å. (1993), 'Explaining Interfirm Cooperation and Innovation: Limits of the Transaction-Cost Approach', in G. Grabher (ed.), *The Embedded Firm: On the Socioeconomics of Industrial Networks*, Routledge, London, pp. 52-64.

Lundvall, B.-Å. and Johnson, B. (1994), 'The Learning Economy', *Journal of Industry Studies*, vol. 1, pp. 23-42.

Maillat, D. (1991), 'The Innovation Process and the Role of the Milieu', in E. Bergman, G. Maier and F. Tödtling (eds), *Regions Reconsidered*, Mansell, London, pp. 103-118.

Nielsen, N. (1994), 'The Concept of a Technological Service Infrastructure', OECD, Paris (mimeo).

Patchell, J. (1993), 'From Production Systems to Learning Systems: Lessons from Japan', *Environment and Planning A*, vol. 25, pp. 797-815.

Pedler, M., Burgoyne, J. and Boydell, T. (1991), *The Learning Company: A Strategy for Sustainable Development*, McGraw-Hill, London.

Perrin, J. (1988), 'New Technologies, Local Synergies and Regional Policies in Europe', in P. Aydalot and D. Keeble (eds), *High Technology Industry and Innovative Environments: The European Experience*, Routledge, London, pp. 139-162.

Porter, M. (1990), *The Competitive Advantage of Nations*, Macmillan, London.

Putnam, R. (1993), *Making Democracy Work*, Princeton University Press, Princeton.

Pyke, F. (1994), *Small Firms, Technical Services and Inter-firm Cooperation*, International Institute for Labour Studies, Geneva.

Reich, R. (1991), *The Work of Nations: Preparing Ourselves for 21st-Century Capitalism*, Knopf, New York.

Rosenfeld, S. (1996), 'Does Cooperation Enhance Competitiveness?: Assessing the Impacts of Inter-Firm Collaboration', *Research Policy*, vol. 25, pp. 247-263.

Smith, K. (1994), 'New Directions in Research and Technology Policy: Identifying the Key Issues', STEP-report, no. 1, The STEP-Group, Oslo.

Smith, K. (1997), 'Economic Infrastructures and Innovation Systems', in C. Edquist (ed.), *Systems of Innovation. Technologies, Institutions and Organizations*, Pinter, London, pp. 86-106.

Storper, M. (1995a), 'Regional Technology Coalitions: An Essential Dimension of National Technology Policy', *Research Policy*, vol. 24, pp. 895-911.

Storper, M. (1995b), 'Competitiveness Policy Options: The Technology–Regions Connection', *Growth and Change*, vol. 26, pp. 285-308.

Storper, M. (1995c), 'The Resurgence of Regional Economies, Ten Years Later: The Region as a Nexus of Untraded Interdependencies', *European Urban and Regional Studies*, vol. 2, pp. 191-221.

Storper, M. and Walker, R. (1989), *The Capitalist Imperative: Territory, Technology, and Industrial Growth*, Basil Blackwell, New York.

Storper, M. and Scott, A.J. (1995), 'The Wealth of Regions', *Futures*, vol. 27, pp. 505-526.

Tödtling, F. (1995), 'Firm Strategies and Restructuring in a Globalised Economy', IIR-Discussion 53, Institute for Urban and Regional Studies, Wirtschaftsuniversität, Vienna.

Tödtling, F. and Sedlacek, S. (1997), 'Regional Economic Transformation and the Innovation System of Styria', *European Planning Studies*, vol. 5, pp. 43-64.

Varaldo, R. and Ferrucci, L (1996), 'The Evolutionary Nature of the Firm within Industrial Districts', *European Planning Studies*, vol. 4, pp. 27-34.

Weinstein, O. (1992), 'High Technology and Flexibility', in P. Cooke, F. Moulaert, E. Swyngedouw, O. Weinstein and P. Wells (eds), *Towards Global Localisation*, UCL Press, London, pp. 19-38.

7 Technological Competitiveness in a Transition Economy: Foreign and Domestic Companies in Hungarian Industry

Györgyi Barta

Introduction

One of the legacies of the state-planned economy is a serious technological gap between developed countries and Eastern Europe. This gap widened at a quickening pace towards the end of the socialist era (i.e. during the 1980s) because it became increasingly impossible to catch up with new Western technologies such as the electronisation of mass consumer products or the use of modern composite materials. The reasons for the technological gap are well-known: Because of the lack of competition, the monopolistic situation of big enterprises, and the permanent shortage of goods and services, there was almost no stimulation of product innovation. A frequently mentioned example is that of the Trabant, a car produced in the German Democratic Republic (GDR – or East Germany) for decades without changing any element of its technology. Because costs of resources were low and unrelated to world-market prices, there were generally no incentives to save energy, raw materials or human resources. For example, the average energy required to produce a $1000 worth of output – in GDP – was 1362 coal units in six East European socialist countries in 1979, while it was 820 coal units in the United Kingdom and 565 in West Germany (Kornai, 1993, p. 319).

The reasons above also explain why the mechanism of technology transfer was so weak in the socialist system. Transfer was totally absent between the more and less developed sectors (for example, between the military industry with its relatively more developed technology and the civil sectors), and the connection was loose between R&D and production. Among the big enterprises in the domestic economy, and among the member states of COMECON, technological transfer was almost completely absent, because both the domestic economy and COMECON were based much more on the division of labour than on competition. Nevertheless, the member-states of COMECON had managed to keep the more or less autarchic character of their own economies. COMECON trade relations followed a star-shaped pattern, with Moscow at the center and little direct trading between other members.

With the collapse of COMECON and the subsequent disintegration of state enterprises, the former connections among the ex-socialist countries and the external and internal connections of, and within, big enterprises disappeared completely. It is still true, however, that the motivation for innovation has not re-appeared, and R&D activity has not been revived yet in the Hungarian economy. But one can now recognise an ability on the part of the new enterprises for a more rapid adjustment to new circumstances.

This chapter discusses the general role and impact of foreign direct investments for advancing technological development in the Hungarian economy, and the gradual formation of a new network of connections between domestic and foreign companies. Next, it analyses the emerging typology of Hungarian companies and describes their systematic integration into the global economy. Some of the integrative processes in the domestic economy can be attributed to the survival or even a renewal of organisational, cooperative and personal connections established in the state-planned economy. Most cases, however, involve the development of new economic links and open up the possibility of technology transfer. Finally, the declining trend in R&D activity is described as a threat to Hungarian competitiveness.

Factors Influencing Technological Development and R&D in the Transitional Years

Peripheral Character and Short-term Survival Strategies

Until 1993, Hungarian industry had undergone a destructive development process. This has involved the collapse of the institutional structure of industry, radical changes in ownership, loss of former markets (domestic as well as foreign), a fall in production and a decrease in both revenues and assets. Using relative prices as the basis of comparison, in 1992 industrial production reached only 54.6 percent of its level in 1989, which is equivalent to the total value of production of Hungarian industry 16 years earlier (Ministry of Industry and Trade, 1995).

Nevertheless, industrial production, exports and productivity started to grow after 1993. The radically restrictive measures of the government in 1995 (affecting, for instance, investments, exports and, especially, domestic consumption) may have slowed down the process of expansion but certainly could not bring it to a halt. The critical disintegrative phase in the trans-formation of the industrial production structure has now come to an end. Change of ownership has advanced significantly: according to the value of paid-in capital (i.e foreign capital paid by acquirers), 29 percent of manufacturing industry remained state property, 37 percent passed into foreign ownership, and 29 percent was acquired by domestic enterprises and private entrepreneurs by 1995. The structure of enterprise organisations has undergone a far-reaching transformation. Large industrial giants, characteristic of the state-planned economy, have disappeared. The share of small enterprises now approximates that of Western European economies. At the same time, the country's industrial organisation has not stabilised yet. Fragmentation is still the prevailing tendency. In general, the new framework of a market-oriented industry is becoming more and more visible (Table 7.1).

An increasingly stable industrial structure and factors contributing to industrial growth point to the fact that the development of Hungarian industry is characterised by expansive tendencies – as generally observed in industries of more developed countries. However, Hungarian firms still occupy only a *peripheral position* in the international economy. 'Eastern Europe has thus been assigned a place in the international economy roughly comparable to what it occupied in earlier centuries: that of a poor cousin in the division of labor with the rest of Europe' (Amsden, Kochanowicz, and Taylor, 1994. p. 5). For example, the relative weight of raw material-intensive and energy-intensive sectors, including metallurgy, has grown.

The most dynamic sectors are dominated by large foreign companies. Examples include Opel, Audi and Suzuki in car manufacturing, IBM and Philips in computer manufacturing, General Electric in lighting-systems production, Unilever, Nestlé and Stollwerck in the food industry, Novartis in pharmaceuticals, Alcoa and Ford in metal processing, and DWA-Voest Alpine

Table 7.1 Characteristics of industrial processes (relative prices, previous year=100)

	1990	1991	1992	1993	1994	1995	1996
GDP	-3.5	-11.9	-3.1	-0.6	+2.9	+1.5	+1.0
Industrial GDP	-7.6	-17.8	-6.7	+3.0	+6.0	+7.3	...
Total investments in the economy	-9.8	-12.3	-1.5	+2.5	+12.3	-5.3	-4.4
Industrial production	-11.1	-24.7	-18.4	+4.7	+9.6	+4.6	+3.4
Industrial export	-17.7	-26.0	-4.7	+3.8	+19.3	+18.5	+18.5
Industrial productivity	-6.2	-8.8	-3.8	+13.4	–	–	...
Industrial investments	-8.4	-2.7	+14.0	-15.0	+7.8	+9.4	...
Industrial production (1985=100)	87.5	69.0	63.3	65.3	71.4	–	...

Source: Central Statistical Office, *Statistical Report on Industry*, 1996, 1997.

in metallurgy. All these companies have either launched joint ventures or established their own enterprises in exclusive ownership. In addition, all of the above-mentioned examples rank among the top 100 companies in Hungary with respect to net sales revenue (*Figyelő*, 1997).

Modernisation of industry, however, is hardly reflected in industrial production or the structure of exports, as shown by the growing importance of subcontracting and contract manufacturing. In these arrangements, the Hungarian producers only manufacture according to a customers' design and specifications. Furthermore, *short-term survival strategies* have usually prevailed in the process of economic transition and this is still the dominant trend today. In other words, companies have pursued a strategy of passive adjustment, restricting production, selling or living off their assets, and placing priority on products with lower added value.

Moderate Incentives for Investment

Economic transformation and restructuring require a considerable supply of capital. The actual demand for capital is, however, hard to assess given that it is determined not only by the available amount of investable capital but also by the absorptive potential of the economy.

The general recession in the Hungarian economy was accompanied by falling investment activity between 1990 and 1993. The value of industrial investments decreased by an estimated 30-35 percent between 1989 and 1993 (taking average international investment rates and real domestic GDP). A significant but only temporary increase of investments was registered in 1994 when the investment rate (total investments as a share of GDP) rose by 20 percent. Due to severely restrictive measures, however, investments fell again in 1995, continuing through 1996. An especially serious decrease in investment was observable in education, health and welfare in the public sector. By contrast, in the private sector, investments receded most notably in transport, postal services and telecommunications (by 10 percent even in 1996). Nevertheless, the second half of 1996 saw a substantial increase in investments in the manufacturing and housing sectors (12 percent and 14 percent, respectively). A year later (1997), figures point to a slow but stable increase which can be attributed to growing exports and to the consolidation of domestic demand.[1]

That economic actors have shown little willingness to invest in the past several years is not due to the unattractive profitability of enterprises. Nor can this behaviour be sufficiently accounted for by the scarcity of investment resources, a consequence of the general lack of capital in the economy. Although these factors have undoubtedly played an important role, processes accompanying *economic transformation* and *shifting markets* have been primarily responsible for the serious cut-backs in industrial investments. They have significantly increased the risks of investment, destabilising the economy and generating a permanently high level of inflation. The same processes have also made the reduction of labour and capital costs more difficult and kept interest rates on loans high. This is the reason why most industrial sectors could not attract the resources needed for development and continued to live off their assets and capital until 1996.

Foreign Direct Investment (FDI) in Hungarian industry

FDI in cash and assets totaled $14 billion in 1995, including investments made before 1989. For many years, the natural capacity of the Hungarian economy to absorb foreign capital was set at about $1.2-$1.5 billion a year. This was also the approximate yearly amount of FDI entering the country, discounting unusually large transactions (Csáki, Sass, and Szalavetz, 1996). By contrast, the number of enterprises with FDI grew fourfold between 1990 and 1995, while the value of paid-in capital was already seven times higher in 1995 than in 1990.

The number of enterprises with FDI increased at a slower rate than the total amount of investments. This implies that the foreign share in newly-established joint ventures increased in the course of time. In addition, the number of companies in exclusive foreign ownership also increased rapidly between 1990 and 1995, their share of the total number of joint ventures with foreign participation rising from 4.1 percent to 43.9 percent. The concentration of foreign capital is high and still increasing. For instance, more than 80 percent of all enterprises with FDI belonged to the category of small-sized companies in 1995, yet more than 90 percent of the value of foreign investments were concentrated in the category of large companies (5.7 percent of all enterprises with FDI).

Although manufacturing has only attracted about 17 percent of the number of foreign joint ventures (Table 7.2), of the total foreign investment in joint ventures, 43 percent has been absorbed by the manufacturing sector and only 12 percent by trade or commerce. It is worth noting that in countries on a comparable level of development, such as Ireland or Portugal, a relatively larger proportion of FDI has been channelled into the financial and real estate sectors.

Table 7.2   Enterprises with FDI by economic sector in 1995

Sectors	Number of enterprises (percent)	Foreign share of equity capital (percent)
Agriculture, forestry, fishing	3.2	1.2
Mining and quarrying	0.3	0.8
Manufacturing	16.7	42.9
Electricity, gas, steam and water supply	0.2	13.2
Construction	4.9	3.5
Trade	48.6	11.9
Hotels and restaurants	4.6	2.5
Transport, storage, communications	3.3	9.0
Financial services	0.6	8.0
Real estate, renting, business activities	14.1	6.4
Education	0.5	0.0
Health and social work	0.7	0.1
Others	2.3	0.5
Total	100.0	100.0

Source: Central Statistical Office, *Hungarian Statistical Yearbook, 1995* (1996, pp. 219-220).

Although the industrial sector was in decline (66 percent of GDP in 1993 vs. 49 percent in 1994), this sector still accounted for more than 40 percent of FDI in 1995. The high proportion of FDI in manufacturing is a result of the particular course of the privatisation process in Hungary. Between 1991 and 1994 the proportion of the fully foreign-owned companies among all enterprises with foreign participation has increased from 15-18 percent to more than 40 percent. This process, however, was to come to an end by 1997. In other words, foreign investment will from now on primarily concentrate on the development of foreign enterprises and joint ventures, as well as on increasing the foreign share in joint ventures, rather than to launch new brownfield and greenfield companies. Nevertheless, as far as the manufacturing sector is concerned, Hungary may well become a major focal point – perhaps even a Central European hub – for FDI as well as for export-oriented investments.

Technology Strategies of Foreign and Domestic Industrial Companies

'In addition to the respective sectoral environment and available technologies, the approach of a given economic organisation to innovation will be principally determined by divergent company strategies' (Tamás, 1995, p. 43). Additionally, the strategies of companies in *domestic and foreign ownership* differ markedly from one another. This has also been illustrated by the fundamental discrepancy between different investment processes described in the previous section. The strategies of companies in domestic and foreign ownership are discussed below.

Technology Strategies of Companies with Foreign Ownership

The favourable conditions of production that foreign investors seek in a country are constituted by an array of factors in the Hungarian case. These include:

- the availability of a relatively inexpensive, large and diverse work force (although there are poorly represented professions as well, especially management, logistics and marketing);
- the willingness and ability of employees to adapt and improve their skills;
- comparably low prices of bankrupt state-owned companies, an adequate infrastructure in certain parts of the country (especially in the capital and western regions);
- a transparent legal environment, a business-friendly atmosphere fostered by the central administration; and
- a stable political situation.

This has motivated the establishment of assembly and production plants (e.g. by Opel and Suzuki in the car industry, Electrolux in the refrigeration industry, General Electric and others in the lighting industry).

Various Hungarian companies have become active in *subcontracting* (for Ford, Audi, etc.). *Contract manufacturing* plays a particularly important role in the footwear, textile and electronics industries. In many cases, investments are aimed at *expanding supply* at the lower end of the consumer market through the manufacture of less sophisticated and cheaper products. (This often entails the continued production of goods that are no longer competitive on the markets of more developed countries. For example, the majority of cars assembled in Hungary belong to the category of cheaper and smaller cars.) Nevertheless, in view of deteriorating conditions of production at home, the decision to invest abroad may sometimes be made in an attempt to 'escape forward' (e.g. in a country such as South Korea where economic growth is slowing down, rigid legal regulations increase labour costs, specific characteristics of the economic organization hinder re-structuring and banks are facing crisis because of the risks of financing domestic investments (*The Economist* May 17, July 12, August 23 1997).

It may be generally concluded, therefore, that the possibility of setting foot in new markets on the one hand and that of re-locating production in a more favourable environment on the other constitute the two most important incentives for foreign investors to invest in Hungary. At the same time, these motivating factors are often difficult to tell from one another. Production is always launched in a foreign country in order to be able to sell at least a part of the goods produced in the host market.

What precisely motivates foreign investors is well indicated by the ways in which they utilise their profits, i.e. whether they choose to reinvest it in the receiving country or to withdraw it from there. Although a great variety of data has been made available on FDI, it is difficult to assess the precise volume of profit repatriation. Nonetheless, reports produced by the Hungarian National Bank indicate that the extent of profit 'export' from Hungary was relatively low until 1995.

> The annual repatriation of profits amounted to about HUF 100 million up until the end of 1994 ($1 was worth approximately HUF 100 in 1994 and is HUF 190 in 1997). This figure went up to HUF 142 million by the end of September 1995. It is hard to estimate the extent withdrawn through other common forms of capital relocation (transfers, capital movements, management contracts, marketing contracts, etc.). As the macroeconomic situation and investment climate is improving, a large-scale withdrawal or flight of foreign capital cannot be expected at present (Csáki,. Sass, and Szalavetz, 1996, p. 9).

Although the *impact of foreign capital on technological standards and innovation* is not unambiguous, it generally tends to be positive. On the one hand, FDI has certainly made a crucial contribution to improving technological standards in Hungary, but this has rarely involved the introduction of high-tech manufacturing procedures. In other words, the improvement of the material conditions of production has usually involved importing technologies that were advanced relative to the former standards of Hungarian industry.

One can identify a marked difference between the technological standards and development of *greenfield* (FDI establishing a new entity) and *brownfield* (i.e. FDI entering through the purchase of Hungarian companies) enterprises. Large investors launching greenfield enterprises have generally applied highly advanced technologies and set strict quality and production standards. Investments of multinational companies (MNCs) have had a significant and generally positive impact on the development of Hungarian industry. Of the 50 most important industrial companies of the world, 35 had entered in the Hungarian economy by 1995. Greenfield investments amounted to 20 percent of all FDI. Among the MNCs investing over 100 million dollars in the Hungarian economy are: General Electric, General Motors, Audi, Ford, Philips, Unilever, Sanofi, Suzuki, Siemens, Guardian Glass, the Prinzhorn group, Sara Lee and Feruzzi. Two-thirds of all greenfield investments were attracted by the engineering industry, especially the car industry, which has absorbed 10 percent of all FDI. Electronics was another important sub-sector, dominated by the investments of Philips (Diczházi, 1995).

By contrast, brownfield investments by foreign companies through the privatisation of state-owned companies are characterised by less advanced technologies, although even those are normally above the former level. One may cite the example of a number of firms in the food and refrigeration industries. There are some examples to the contrary as well, such as Graboplast, an Austro-Hungarian joint venture offering quality products that meet the highest international standards (Grayson and Bodily, 1996).

The situation is no different with medium-sized enterprises where FDI was realised by means of an increase in capital. Technological standards at such enterprises can also be said to be inferior to high-tech manufacturing processes employed at leading companies. Nonetheless, even the import of such second-rate technologies may help to reduce the serious technological handicap of Hungarian industry.

Technology Strategies of Domestic Companies

Distinct technology strategies can be identified in domestic companies. For the purposes of the following discussion, I introduce a typology:

*   companies concentrating on the domestic market;
*   companies integrated into regional networks: industrial districts;
*   companies integrated into high-tech networks: technological areas; and
*   companies integrated into the new international division of labour.

Companies Concentrating on the Domestic Market   In Hungary, the number of companies doubled between 1990 and 1996. As many as 750,000 sole proprietorships were registered in 1996, in addition to 284,000 joint enterprises, 126,000 of which were and 158,000 of which were *not* listed as a legal entity. No initial capital investment is required in case of sole proprietorships and joint enterprises not operating as legal entities. More important companies, including companies in foreign ownership, however, all function as legal entities. Financial necessity has forced the majority of entrepreneurs to launch private enterprises, nearly all of which are small: 99 percent of the joint enterprises not registered as a legal entity employ less than 10 people. Most of these entrepreneurs are in effect self-employed, not capable of either accumulating capital or expanding their enterprise. They usually run their enterprises on a day-to-day basis living on their revenues and assets. As a rule, this category figures prominently in underdeveloped economies with considerable unemployment. As much as 22 percent of the active population can be ascribed to this group in Hungary. Similar rates have been recorded in Ireland, Italy, Spain and Portugal, and their proportion runs even higher (35 percent) in Greece (Laky, 1995).

Needless to say, a number of enterprises are in a somewhat better situation. While not manufacturing their products exclusively for local markets, these small enterprises are also far from being able to conceive of exporting their products abroad. In general, small enterprises do not develop into medium-sized or large companies. Our field studies carried out in Budapest have found that some entrepreneurs *deliberately* aim at keeping their enterprises small. The large number of small enterprises is, therefore, not only a function of economic exigencies. Not surprisingly, there is considerable fluctuation or turbulence among small enterprises.

Both the economic situation and the dominant entrepreneurial attitude are responsible for a low level of technology. Other empirical studies have indicated, for example, that: 'Hungarian greenfield companies, nearly all of which can be labelled SMEs, show a strong tendency towards technological, product and process stagnation' (Makó, Ellingstad, and Kuczi, 1997, p. 21) and that 'this type of economic organisation [i.e. small enterprises] is the most common among domestic companies. This corresponds to the model of peripheral development and can also be seen as a reaction to the overcentralisation of industrial structure' (Tamás, 1995, p. 44).

Companies Integrated into Regional Networks: Industrial Districts
Hungarian industry is almost completely devoid of networks of small-sized

companies. These networks, which are so characteristic in some other economies, are based typically on a long tradition of craftsmanship and family enterprises. Nevertheless, rudimentary forms of industrial regions have already appeared. For instance, in a village near Budapest, about twenty carpenters have decided to cooperate and take on orders from the capital together, instead of competing with one another (Kuczi, 1993).

Companies Integrated into High-Tech Networks: Technological Areas
There are a few examples of high-tech networks in Hungarian industry, mostly in the economically dynamic agglomeration of Budapest and the region lying between Budapest and the Austrian border. A good example is provided by the industrial development of Székesfehérvár, a city situated 60 km to the west of Budapest, which attracts the highest amount of FDI after Budapest. Factories and branch plants have been settled here by ALCOA, IBM, Bericap, Emerson Electric, Fisher Rosemount, Texas Instruments, Kenwood, Ford, Nokia, Philips, Parmalat, Stollwerck, Loranger, Shell Gas, and Denso.

Székesfehérvár, an attractive medium-sized town rich in historic monuments (having functioned as a capital in the early days of Hungarian history), has a population of 100,000 people. Its success story is to be attributed to the simultaneous presence of a number of factors. It has inherited a developed industrial environment and a skilled, comparably young workforce from the state-planned economy (Székesfehérvár was an industrial centre for aluminium processing, bus and television manufacturing). This legacy is in part responsible for its outstanding infrastructure, such as a direct motorway leading to the capital. The town's geographical location and its proximity to the capital has become even more valuable after the transition. These circumstances have been accompanied, however, by a series of equally important factors, including a carefully devised economic policy and the local cooperation of political and economic leadership. It has been widely recognized that FDI is crucial to the development of the local economy. Industrial parks were established that were primarily occupied by foreign companies.

The Industrial Park of Videoton needs mentioning in particular: a model example of the way in which industrial parks can be used to attract foreign capital. (Foreign companies settling here have engaged in cooperative activities with Videoton, a Hungarian company producing spare parts for electronic appliances and telecommunication. Videoton has not only become a cooperative subcontracting and outsourcing partner, but is also providing whatever economic services are required by these foreign companies. Consequently, in addition to having helped foreign companies to settle there, Videoton is continuously supporting their production as well.)

The network of connections developed by these companies and their participation in the spatial division of labour were the focus of a case study described below.

Areas of Production Input   Rather than keeping hold of market areas and settlements for their products and services, Székesfehérvár's traditional companies have been much more interested in preserving their circle of suppliers of raw materials, energy, semi-finished products or, more precisely, in clinging to the same areas of origin for these production inputs. The geographical origin of these production inputs may greatly vary depending on the company and sector. Most raw materials, however, come from Eastern Europe and countries of the former Soviet Union. The branch plants of most Hungarian suppliers are also to be found outside the boundaries of Székesfehérvár and its region, although shipping costs have already started to play a role in choosing between branch plants today.

A significant difference could be observed between domestic and foreign companies with regard to the geographical location of their suppliers. Hungarian companies rely to a much higher extent on inputs originating from the same region, or from Hungary in general, than foreign companies. Furthermore, marked discrepancies were found between greenfield enterprises and privatised companies (Table 7.3).

This empirical research by Makó, Ellingstad, and Kuczi (1997) shows that older privatised companies receive a larger part of their inputs from within the region and in particular from Hungary as a whole than do new greenfield companies. By the same token, it is hardly surprising that a larger part of the inputs of foreign enterprises in Hungary comes from abroad (mostly from countries of the European Union) rather than from domestic companies. This shows that the development of connections between companies takes a long time. The selection of partners depends largely on the knowledge of the local environment, mutual trust and familiarity.

Market Areas   The difference is no less striking in the case of market areas (Table 7.4). Hungarian companies sell their products within the Székesfehérvár region and within Hungary at a much higher rate than foreign companies do. The same difference can be noted with regard to greenfield and privatised enterprises in foreign ownership. Similarly to the regional distribution of production inputs, these tendencies are to be attributed to the time required for

Table 7.3   **Distribution of production inputs by area of origin in 1997 (percent)**

	Region	Hungary	EU	non-EU countries
Hungarian companies				
Greenfield companies	30.3	29.2	13.8	4.4
Privatised companies	27.2	50.3	15.1	4.9
Total	27.8	46.4	14.8	4.8
Foreign companies				
Greenfield companies	7.1	15.1	63.5	14.5
Privatised companies	22.5	54.3	20.8	2.5
Total	9.3	21.8	54.9	14.0

Source: Makó, Ellingstad, and Kuczi (1997, p. 8).

developing and shifting relationships between companies. Foreign companies, especially those without a tradition in the region, are only loosely bound to the regional economy.

The system of company relations plays an important role in the economic development of a region. Developing and expanding connections between companies promote the integration of foreign companies into the domestic economy. They foster the encounter of different 'cultures' of production and may help companies with lower technological standards to overcome their handicap. It can be argued that this is one of the most crucial functions of foreign direct investment in the economy of the receiving country (Fromhold-Eisebith, Chapter 9, *this volume*).

Székesfehérvár's industrial region is by no means an isolated area. It has especially close connections to Budapest and the Budapest agglomeration. It is most likely that Budapest's proximity represents an advantage rather than a hindrance for the Székesfehérvár region. While Budapest does not draw away the region's growth potential, foreign capital settling there is in part attracted precisely by the vicinity of the capital. Budapest and its agglomeration may often constitute a market for companies in Székesfehérvár. It is to be added, however, that the spatial division of labour is still not totally balanced and some important prerequisites of production are still missing in Székesfehérvár. The general availability and standards of local higher education, for instance,

Table 7.4 **Regional tendencies in the distribution of output in 1997 (percent)**

Region	Hungary	EU	non-EU countries
Hungarian companies			
Greenfield companies	58.1	37.5	3.8	0.6
Privatised companies	43.4	38.9	11.0	6.8
Total	45.8	38.7	9.8	5.8
Foreign companies			
Greenfield companies	15.7	22.8	49.6	12.0
Privatised companies	30.0	47.3	16.3	6.5
Total	16.7	25.0	43.7	10.8

Source: Makó, Ellingstad, and Kuczi (1997, p. 8).

remain unsatisfactory and institutions and facilities for research and development are missing. Consequently, managers and the most skilled workforce still commute to the region from Budapest.

Hungarian Companies Integrated into the New International Division of Labour Hungarian companies may perform quite different functions in the international division of labour. I will use the following examples to illustrate the four most characteristic of these:

(1) Manufacture of Hungarian products of mediocre quality for the markets of developed countries where the manufacture of these products has been cut back or stopped altogether.
(2) Renewed expansion into Eastern European markets of former COMECON countries.
(3) Subcontracting and supply of semi-finished products for usually larger foreign companies producing in Hungary.
(4) A growing share of contract manufacturing.

The *first* role is perhaps best exemplified by the recent history of the Hungarian steel industry (Barta and Poszmik, 1997). The technological development of the industry was delayed and remained incomplete, as was true of other Eastern European countries. Given the availability of cheap labour and ever-decreasing yet significant state subsidies, however, the prices of Hungarian steel products could be kept below world market prices. Thus Hungarian steel works could boast impressive export figures despite high production costs and outdated technologies. The obsolescence of the technologies used is underscored by the fact that the share of converter processes is still the highest in Hungary, reaching 87 percent in 1992 and 95 percent in 1996. While this technology had already been completely abandoned in developed European countries in the 1980s, as many as 42 percent of all Hungarian steel products were still manufactured using this technology in 1990. However, it is now the case that developed countries, especially Italy and Germany within the EU, have clearly become the main markets for Hungarian steel exports (Table 7.5).

With regard to the *second* role, Hungarian industrial companies have shown a growing interest in recovering their Eastern European markets lost in the wake of the political-economic transition. This tendency is illustrated by strategies of firms in the Hungarian pharmaceutical industry. This sector was one of the most important suppliers of medicines to the COMECON in the state-planned economy. Countries of the former Soviet Union and other Eastern European countries were forced to reduce their imports of medicines dramatically due to the general economic crisis after the transition. A gradual improvement of the situation could be observed by 1997: The economies of Eastern European countries have more or less recovered, social security systems continue to function and solvent demand is present. At the same time, there is keen competition and consequent protectionism in the region. The most successful way of securing markets for Hungarian products seems to be the purchase or establishment of manufacturing plants in Eastern Europe.

The situation is quite different in countries of the former Soviet Union. (Social security systems normally keep a considerable part of domestic pharmaceutical production alive by generating constant demand for its products). Welfare systems of countries of the former Soviet Union, however, do not appear to be viable, consequently, selling products requires intense marketing activities in local markets there. The largest Hungarian exporters, i.e. Richter, Chinoin and Egis, have nevertheless recovered their former positions. Given increasing protectionism in these countries as well, further expansion can only be achieved by Hungarian companies that also establish local branch plants. The first such enterprise will be launched by Richter in Russia this year (Antalóczy, 19 ˊ).

Table 7.5   Distribution of iron and steel exports among destination
countries between 1992 and 1995 (percent)

Countries	1992	1993	1994	1995
All transitional and non-market countries	17.0	23.8	7.6	10.6
Only Eastern European countries	14.0	7.1	5.0	8.0
All market economies	83.0	74.9	90.3	87.2
Developed countries	62.9	50.1	75.9	76.3
Only EU	44.8	31.4	57.9	69.4
Only EFTA	9.1	6.5	5.8	0.5
Only outside Europe	5.5	8.3	11.6	4.0
Developing countries	20.0	24.8	14.4	10.9
Other	0.0	1.3	2.1	2.2
Total	100.0	100.0	100.0	100.0

Source: Central Statistical Office, *Statistical Yearbook on Foreign Trade, 1992, 1993, 1994, 1995.*

The third function, the evolution of a domestic *system of subcontractors and suppliers*, is a slow process. Developing contacts takes time, and foreign companies settling in Hungary encounter difficulties in finding Hungarian subcontractors. Technological standards of small and medium-sized domestic companies often fail to meet the quality requirements of large foreign companies and multinationals producing in Hungary.

Suzuki is a frequently-cited example. Hungarian Suzuki, one of the biggest foreign investments in Hungary, is interested in trying to cooperate with Hungarian industry. According to international conventions on car exports, Suzuki can only increase its sales in Western European markets if at least 60 percent of its final product is manufactured in Hungary or Europe. About 400 Hungarian enterprises have answered Suzuki's call for subcontracting, but only 34 were able to pass Suzuki's strict quality tests. Nonetheless, it has been found that the system of cooperation between the Hungarian and Japanese partners, even if it is not always as smooth as it could be, is workable.

Moreover, since Hungarian subcontractors have to rely in turn on other subcontractors, the high quality standards of the Japanese company are spread to other areas of Hungarian industry as well. At any rate, the network of subcontractors is now complete, the Japanese firm has signaled no intention of wanting to expand it or alter it in any significant way.

Generally, data on subcontracting relations are scarce, but indications from company expenditures indicate that an increasing share of material costs has been taken up by subcontracting. Various private and state organisations are engaged in supporting, developing and referring subcontracting.

Contract manufacturing had already appeared in Hungarian industry before the transition, mostly owing to the economic reforms of 1968. Products manufactured under manufacturing contracts obviously yield smaller profits than brand-name goods. Manufacturing contracts with multinational companies or other important foreign partners with products of the highest international quality, however, mean not only reasonable profits but they also guarantee high technological standards for domestic enterprises. Videoton in Székesfehérvár, for example, has begun contract manufacturing for Nokia, Ericsson, Eltek, Alcatel, Samsung and others.

The share of contract manufacturing production has risen notably in Hungarian industry (Table 7.6). Such manufactured products amounted to 30.1 percent of all Hungarian industrial exports in 1995. The share is as high as 64.7 percent in the lighting industry, 95 percent in the clothing industry and 97 percent in the footwear industry in 1995 (Cseh, 1997).

Contract manufacturing has rescued the clothing industry in Hungary. In general, it can be said that subcontracted manufacturing can be accepted as a *survival* strategy. One of its advantages is that a number of former out-sourcing companies have decided to start their own enterprises in Hungary (e.g. Levi-Strauss, Elit Group from the UK, Baumler and Rosner from Germany). Subcontracting, however, hinders further growth and progress. Companies with contract manufacturing contracts often become complacent and ignore the need for independent development. In addition, they cannot accumulate sufficient amounts of capital to finance their own undertakings. In many instances, contract-manufacturing companies have to rely on outside support for launching independent enterprises. Various schemes, such as the PHARE programme and the Hungarian Investment and Trade Development Company-ITD Hungary, provide aid in such cases.

Table 7.6   Sectoral changes in subcontracted manufactured exports

Sector	1991		1995	
	Total exports (US$ million)	Share of subcontracted manufactured exports (percent)	Total exports (US$ million)	Share of subcontracted manufactured Exports (percent)
Mining	16	0	13	7.7
Electrical energy	–	–	5	–
Metallurgy	910	5.9	1142	14.4
Engineering	2566	18.1	3889	28.5
Construction	21	0	13	7.7
Construction materials.	217	0	236	0.8
Chemicals	1775	19.0	2191	9.1
Lighting	1672	52.7	2285	64.7
Other industries	52	13.5	82	20.7
Total industry	7229	24.1	9856	30.1
Foreign trade	9972	17.9	12861	24.1

Source: Cseh (1997, p. 21).

Research and Development (R&D)

Relatively large research institutes and facilities were maintained in the state-planned economy. They operated, however, more or less in isolation from economic organisations and remained inefficient as a consequence. Institutes subordinated to ministries of a given industrial sector were normally responsible for innovation and R&D. The comparably large contribution of the Hungarian Academy of Sciences to basic and applied research constituted another important feature of R&D in the state-planned economy. Finally, it should be noted that the old system of R&D was based on a linear model coupled with supply-oriented innovation policies of the state. The state financed all R&D and educational activities in the public sector. Although the centralised system of planning had many weaknesses, it can be argued that it has invariably guaranteed a reasonably high level of resources for the purposes of research, development and other innovative activities (Table 7.7) (van Geenhuizen, 1997).

> In Hungary innovation activity declined in absolute terms as well, not only becoming less and less significant in comparison with the highly industrialized countries, but year after year losing its strength in comparison with its own earlier performance (Szántó, 1994).

Between 1988 and 1993, absolute expenditures on R&D decreased by 54 percent, falling at a much faster rate than GDP. The share of expenditures on R&D in Hungary is in any case quite low in international comparison: 2 percent of the GDP in 1989, falling to 1 percent in 1993. The number of scientific researchers had decreased from 17,550 to 11,820 by 1993. The pressure for further cut-backs is still intense: The president of the Hungarian Academy of Sciences announced in 1997 that another 25 percent of the 4,000 researchers employed by the Academy will be made redundant in 1998.

The increasing scarcity of R&D funds has affected the various types of organisations in a differentiated way. Some industrial research institutes have been closed down, while others just barely survive. Between 1988 and 1993, the nominal value of the revenues received by research institutes decreased by 82 percent. R&D activities as well as the underlying institutional system have been subject to a fundamental restructuring process. While independent R&D activities and separate R&D departments within companies have been on the

Table 7.7   The R&D system in state-planned economies and market economies

	State-planned economies	Market economies
Technology Supply	In isolated institutions	In enterprises, enterprise networks and other organisations; priority to rapidly adapting R&D teams.
	Technology-led (imitative) R&D	Technology and market-led R&D
	Cost and price insensitive	Market-oriented services
	No technology transfer	Technology transfer playing an important role (interchange)
	Linear innovation model	Interactive model
Technology Demand	Determined by central planning	Determined by available technologies and the market
	Fixed	Flexible
	Disciplinary approach	Priority to interdisciplinary approach

Source: van Geenhuizen (1997, p. 8).

decline, the number of occasional and specialised R&D tasks has increased significantly.

Industrial enterprises have generally shown less and less interest in innovation. Spending on R&D has decreased. Enterprises have to rely on their own resources to finance the bulk of their R&D costs. The government has only limited funds available to foster domestic R&D potential, and foreign investors are rarely interested in sponsoring R&D in Hungary. Consequently, R&D activities in foreign enterprises are not pursued with great intensity (Tamás,

1995). It must be noted, however, that the dwindling of R&D is not only a result of changing spending policies. It seems to be overwhelmingly the case that company managements pay much more attention to foreign technologies than to the potential results of in-house research. This explains the growing import of licences and know-how, especially in joint ventures with foreign participation.

The cut-backs in R&D activities are also to be attributed to the fact that, on average, companies are getting smaller. As already mentioned, one of the principal reasons for the fragmentation of the structure of enterprise organisations is the disintegration of large industrial companies dominating the state-planned economy. These had functioned as the main centres of technological development. Small companies are less innovative and their financial possibilities are, of course, much more limited.

It is only the Budapest agglomeration and the north-west part of the country that appear at present to have a prospect of developing into areas with a significant concentration of high-tech companies. Multinationals and large foreign companies could play a leading role in creating industrial and technological parks there.

It is clear that the restructuring of Hungarian economy has had a significant influence on R&D activities. One obvious consequence of the transition was the partial withdrawal of the State from organizing and financing R&D. At the same time State support cannot be replaced by unfettered market mechanisms in this area (Szántó, 1994, p. 609). Government action, however, will always have to be tailored to market conditions. Consequently, successful R&D activity and an effective relationship between R&D and production can only be expected after the transitional period has come to an end in East European countries.

Conclusion

It can safely be concluded that the uncompromising restrictive policy package of the Hungarian government (this has included the drastic reduction of domestic consumption, significant increase of centrally regulated prices, devaluation of the Hungarian currency, and cut-backs in imports and welfare programs) in the 1990s has been successful. Even more importantly, however, certain changes have taken place in the economy which are now beginning to yield results. The national economy has started to grow, and the GDP was expected to increase by 2-3 percent in 1997 and 3-4 percent in 1998. Moreover, this growth will probably not upset the relative economic equilibrium (for example, a 7.5 percent rise in industrial production will have been fully absorbed by exports in 1997).

At the same time, export-oriented growth is generated by a relatively small segment of the economy. A handful of sectors and a handful of companies provide the substantial share of expanding production. Out of 28 industrial sectors, only 7 sectors have shown an above-average increase in production, including the machine, wood, printing and paper industries. Two basic factors determine the differentiation of companies: participation in exports and ownership – more precisely foreign or domestic ownership.

Some argue (cf. Budapest Bank, 1997) that the organizational transformation of Hungarian economy has been characterized by a bipolar industrial structure. Companies and sectors that have been successfully integrated into transnational networks and boast rapid growth represent one extreme. These enterprises are for the most part owned by multinational companies. It is estimated that this group of companies is responsible for 60-70 percent of overall industrial growth. The number of small and medium-sized enterprises linked to MNCs has slowly started to increase as well, especially in car manufacturing, electronics, the paper and packaging industries. By contrast, the majority of firms are located at the opposite 'pole'. Their production is either stagnant or increasing only at a very slow rate, and their technological innovations are highly restricted. These companies manufacture only for a small domestic market. (A similar classification of firms divides enterprises into non-export-oriented, assembling-exporting and export-oriented domestic companies (Éltető and Sass, 1997).)

Several case studies and research projects have demonstrated that the integration of foreign companies in the Hungarian economy, those of multinational companies in particular, is a very slow process. At the same time, it would be inappropriate to characterize these enterprises as isolated enclaves. Nevertheless, connections between companies representing the two poles of Hungarian economy, i.e. enterprises producing for the domestic market, mostly in Hungarian ownership and multinational companies together with their subcontractors, are quite limited. Furthermore, subcontractors of multinational companies are often foreign companies themselves; domestic enterprises play a fairly minor role in the new division of labour (with manufacturing activities requiring low technological standards and an unskilled, cheap workforce).

These are the general processes that determine overall technological standards in the Hungarian economy as well as its relative position in comparison to leading economies. There is no doubt that the initial emergence of foreign direct investment as well as its by now enormous contribution to production have had a highly stimulating effect on Hungarian economy. Economic growth and the expansion of exports is mostly due to this set of companies. Their technological standards often involve the actual import of the latest manufacturing procedures, but are in any case more advanced than the Hungarian average. These standards, however, are spreading very slowly through the economy. This is primarily explained by the fact that an increasing differentiation of companies and a peculiar division of labour characterizes the domestic economy. In other words, even though large foreign companies are rightly regarded as 'motors' of the Hungarian economy, the slow diffusion of higher technological standards is to be attributed to their delayed integration in the domestic economic environment.

Note

1 These forecasts are always to be treated with caution. The fluctuation of investment rates is closely linked to changes in the political environment, or, more precisely, to the 4-year cycle of general elections. Since 1994 was the last year of the previous government's term, it is not surprising that an economic policy of artificial revitalisation was adopted by the outgoing government in the hope of reversing negative economic trends of the preceding years. The next general elections take place in 1998.

References

Amsden, A.H., Kochanowicz, J. and Taylor, L. (1994), *The Market Meets Its Match: Restructuring the Economies of Eastern Europe*, Harvard University Press, Cambridge, MA.
Antalóczy, K. (1997), A magyar gyógyszeripar versenyképessége (Competitiveness of the Hungarian Pharmaceutical Industry), Műhelytanulmányok, 17, Budapesti Közgazdaságtudományi Egyetem, Vállalatgazdaságtan tanszék, p. 51.
Barta, Gy. (1994), 'Foreign Investment in the Hungarian Economy: The Role of Transnational Companies', in P. Dicken and M. Quévit (eds), *Transnational Corporations and European Regional Restructuring*, Netherlands Geographical Studies, no. 181, Utrecht, pp. 131-151.
Barta, Gy. (1997), 'Foreign Direct Investment and the Modernisation of the Hungarian Economy', Manuscript, Budapest, 21 pp.

Barta, Gy. (1997), *Ipari nagyvállalatok Fejér megyében* (Big Enterprises in the Industry of Fejér County), Study for the regional development plan of Fejér county, Centre for Regional Studies, Budapest, 24 pp.

Barta, Gy., Králik, M. and Perger, É. (1997), 'Achievements and Conflicts of Modernisation in Hungary', *European Spatial Research and Policy*, vol. 4, no. 2, pp. 61-83.

Barta, Gy. and Poszmik, E. (1997), *A vas-és acélipar versenyképességét befolyásoló tényezők* (Competitiveness in the Hungarian Iron and Steel Industry), Műhelytanulmányok, 15, Budapesti Közgazdaságtudományi Egyetem, Vállalatgazdaságtan tanszék, p. 73.

Budapest Bank (1997), *Makrogazdasági elemzések, prognózis 1997-re* (Macroeconomic Analysis and Prognosis for 1997), Budapest Bank Rt., Budapest, March.

Central Statistical Office (1996) *Hungarian Statistical Yearbook*, 1995, KSH Iparstatisztikai jelentés (Central Statistical Office), Budapest.

Central Statistical Office (various years), *Statistical Report on Industry*, KSH Iparstatisztikai jelentés (Central Statistical Office), Budapest.

Central Statistical Office (various years) *Statistical Report on Foreign Trade*, KSH Iparstatisztikai jelentés (Central Statistical Office), Budapest.

Csáki, Gy., Sass, M. and Szalavetz, A. (1996), 'Reinforcing the Modernisation Role of Foreign Direct Investment in Hungary', Working Paper no. 62, Institute for World Economics, Hungarian Academy of Sciences, Budapest.

Cseh, J. (1997), A magyar textil-és textilruházati ipar helyzete, a versenyképességét meghatározó tényezők. (Competitiveness of the Hungarian Textile and Clothing Industries), Műhelytanulmányok, 13, Budapesti Közgazdaságtudományi Egyetem, Vállalatgazdaságtan tanszék. p. 76.

Diczházi, B. (1996), Külföldi beruházások Magyarországon 1995 végéig. Külföldi, zöldmezős ipari beruházások Magyarországon. (FDI in Hungary until the End of 1995: Foreign Greenfield Investments in the Industry in Hungary), Tulajdon Alapítvány, Privatizációs Kutatóintézet, Budapest, p. 10.

Éltető, A. and Sass, M. (1997): A külföldi befektetők döntéseit és a vállalati működést befolyásoló tényező Magyarországon az exporttevékenység tükrében, (Foreign Ownership and Export Activity as the Most Important Determinants of Economic Performance in Hungarian Companies), Közgazdasági Szemle, június, pp. 531-546.

Figyelő (1997), '"TOP 200": The Largest Hungarian Companies, 1997', Supplement.

Grayson, L.E. and Bodily, S.E. (1996), *Integration into the World Economy: Companies in Transition in the Czech Republic, Slovakia, and Hungary*, IIASA, Laxenburg.

Kornai, J. (1993), *A szocialista rendszer: Kritikai politikai gazdaságtan* (The Socialist System: The Political Economy of Communism), Heti Világgazdasági Kiadói RT., Budapest.

Kuczi, T. (1993), 'Collective Work Organization of Small Firms in Hungary', Conference on the Social Embeddedness of the Economic Transformation in Central and Eastern Europe, Social Science Research Centre Berlin (WZB), Sept. 24-25, 1993, p. 21.

Laky, T. (1995), A magángazdaság kialakulása és a foglalkoztatottság (Private economy and employment), *Közgazdasági Szemle*, 7-8, pp. 685-709.

Makó, Cs., Ellingstad, M. and Kuczi, T. (1997), 'REGIS: Székesfehérvár Region – Survey Results and Interpretation', Centre for Social Conflict Research, Hungarian Academy of Sciences, Budapest, p. 35.

Ministry of Industry and Trade (1995), *Középtávú iparpolitika az ipar versenyképességének növeléséért* (A report about the medium-range industrial policy for increasing competitiveness), Ipari és Kereskedelmi Minisztérium (911/795/1995), Ministry of Industry and Trade, Budapest, December.

Szántó, B. (1994), 'Innovation in Crisis: Hungary Before and After the Watershed of 1989', *Technovation*, vol. 14, pp. 601-611.

Tamás, P. (1995), *Innovációs folyamatok a magyar gazdaságban* (*Innovation in the Hungarian Economy*), Országos Műszaki Fejlesztési Bizottság, Budapest.

van Geenhuizen, M.. (1997), 'Opportunities for Innovation in Central and Eastern Europe: The Role of Foreign Direct Investment', Paper for the RSA Nederland Dag 1997 'Innovatie in bedrijf en regio: strategie en praktijk', Utrecht, April 11, 1997.

8 In Search of Innovativeness: The Case of Zhong'guancun

Jici Wang

Introduction

China's economy is racing forward at an impressive rate, but it has been acknowledged that if it wants to maintain its economic growth momentum and sustain competitiveness, great effort should be made to boost science and technology as well as education. As part of that effort, the competitiveness of the local innovative milieu has become essential. This chapter analyzes the process of creating innovative capabilities in Zhong'guancun, the most knowledge-intensive region in China. The need to search for more innovative capabilities as part of an appropriate long-term strategy has been newly discussed by the Zhong'guancun local government. This chapter paints a picture of the challenge faced by local policymakers, and discusses the potential role of local network formation in particular. The question posed, more precisely, is: What are the tasks local policy-makers in Zhong'guancun are faced with in the effort to create a local innovative milieu, capable of responding to exogenous changes caused by globalization? The chapter identifies weaknesses in the local innovative milieu of Zhong'guancun, and analyzes the current policy approach to enhancing local innovative networks. It stresses the need for local networks and interaction, but also that external resources are needed for successful innovation.

The paper is divided into four sections. The first section outlines problems related to China's high-technology (high-tech) development. The second section introduces the Zhong'guancun region, and examines the prospects of developing it into an innovative environment. The third section identifies the weaknesses of the local innovative milieu. The fourth section suggests a policy to enhance local innovative networks.

Fundamental Dilemmas in China's High-Tech Development

During the twenty years since China began its economic reform and opened its door to the outside world in 1978, tremendous changes have taken place. It is clear that various efforts by the Chinese to attract foreign know-how and expertise have helped to modernize the country substantially. However, the excessive influx of foreign products has caused apprehension about the ability of domestic sectors to grow in the face of competition from foreign products. China's growing participation in the world economy has left the leadership confronted with a fundamental dilemma. Especially in high-tech sectors, there is a difficult task to balance the trade-offs between the independence of innovative action and reaping the benefits that could come from the technological advancement of developed countries.

China is viewed by multinational corporations (MNCs) as only a minute source of high-tech products but a huge and profitable market for exploitation by their own products. They have also noticed that Chinese science and technology personnel has impressive R&D potential. While the need of MNCs to seek out low-cost labor manufacturing sites has declined, the need to seek out sources of R&D has grown. Two facts speak for themselves. One is the increasing amount of foreign capital in China's high-tech zones – invested not only in branch assembly plants for high-tech products, but also in R&D centers for tailoring products designed originally for customers in other countries to meet the needs of the Chinese market. The second fact is that a growing number of international high-tech companies are establishing their branch organizations not only in Chinese export processing zones, but also in its science parks.[1]

The following paragraphs of a 1996 report by United Nations Industrial Development Organization (UNIDO) took a grave view of the matter:

> The rapid pace of technological development and innovation and the establishment of new organizational structures in developed countries pose a formidable challenge to developing countries and economies in transition in their efforts to develop new technological capabilities, to innovate and to ensure sustained industrial growth. The emergence of generic technologies … has significantly changed the nature and scope of industrial competitiveness and the organizational structures at the firm level. At the same time, rapid technological development has further concentrated innovative capabilities in the industrialized countries, leading to a growing gap in such capabilities between industrialized and developing countries …

Many developing countries, and especially LDCs, have only a small manufacturing base and have not yet established a position in international markets for manufactures ... These countries would then have to adapt to new forms of competition arising from rapid technological change, because low wages will no longer guarantee success in sustaining competitiveness. The acquisition of innovative capabilities is becoming essential, for somewhere in the world someone will almost certainly be developing innovative technology that will pose a competitive threat to existing producers (UNIDO, 1996, pp. 4 and 10).

These points are firmly buttressed when they are seen from a Chinese perspective. At the present time, Chinese high-tech firms are concerned by a growing number of prestigious MNCs establishing themselves in Chinese knowledge-intensive regions. There is a wide technological gap between China and developed countries. However, the words of the noted Chinese atomic nuclear physicist Qian Sanqiang express the national spirit: 'The Chinese are not stupid at all. What foreigners can do, the Chinese, through hard work, can accomplish as well' (Qian, 1991, p. 40). This hard work is aimed toward two objectives: first, to raise the quality of labor-intensive products and, second, to develop high-tech products in order to improve the nation's competitive advantage.

China's high-tech development has been nagged by the shortage of capital and skilled labor. The amount of money China spends annually on science and technology research accounts for a meager 0.5 percent of its Gross Domestic Product (GDP), compared with the 2 to 3 percent of GDP in developed countries (Si, 1997). Skilled labor shortages are particularly evident in the rapidly growing cities, due to the low percentage of the population receiving tertiary education and continuing inter-regional and inter-firm restraints on migration. These shortcomings arose from the economic growth strategy of relying mainly on building more production facilities to attract foreign capital and technology rather than advancing indigenous Chinese technology.

China's future economic success lies in finding the best combination of domestically- and foreign-generated technologies, and blending these to its advantage. China is now caught up in the paradox that faces all peripheral countries: to develop its own high-tech sectors empowered by foreign capital without releasing its technology to potential competitors of the world. The most important task facing Chinese policy-makers, besides the attraction of foreign investment into high-tech sector, is to create local innovative environments, both for bridging the gap between domestic research and production and for developing its own high technology. This has not been spelled out yet in clear terms by Beijing.

There is good reason to believe that certain pockets of excellence concerning innovative capabilities have been created since the end of the 1970s: The Chinese government has actively issued a set of programs and measures to promote the development of high-tech industry under multifaceted reforms, especially an effective reform of the scientific and technological system. The milieu created has nurtured a number of vigorous high- and new-tech firms and made them achieve amazing success. However, the potential of research institutes and firms to explore their innovative capability generally is far from fully developed. According to statistics from the National Education Commission, the potential for technology research links between universities and industries has been exaggerated. Of the scientific research findings gained in the institutions of higher learning during the early 1990s, only 30 percent have found practical application in production and only 10 percent have achieved measurable economic advantages. Moreover, 85 percent of industrial enterprises now believe that the scientific research findings made by research institutions remain at a relatively low level (Sui, 1992).

International experience demonstrates that the process of innovation is highly complex. It incorporates feedback from the market to production, engineering, and design, and it requires integration of a firm's R&D and manufacturing with related activities of suppliers and customers. It has been noted that regional/local innovation networks that are based on long-term co-operative relations between businesses, politico-administrative authorities and scientific institutions are essential to overcome the barriers to innovation in a systematic way (e.g., Braczyk and Heidenreich, 1998). In comparison to such Western experiences, the Chinese local system of innovation has some unique features and cannot be understood fully within only a Western theoretical context. In the Chinese industrial organisation, the tendencies of both vertical integration and networking arrangements seem to be manifested. In light of Western experiences only, this might be interpreted as a lack of real awareness of how high technology develops, as well as a lack of sufficient change in management and organization. However, the situation can also be interpreted as one that reflects the developmental stage of the Chinese transition economy.

Zhong'guancun's Evolving Innovative Environment

The Zhong'guancun region is situated in a western suburb of metropolitan Beijing and the south-eastern part of the Administrative District of Haidian. With a radius of about 1.3 kilometers, it has long been renowned as China's largest knowledge-intensive region where important state research and education establishments are located. In this biggest 'brain concentration', outstanding professionals in research institutes and universities have been engaged in national planned high-tech projects. Since the early 1980s the region has become one of the notable gold mines of new-tech products at the state level. While in other parts of Beijing mass production is in crisis – plant closures and layoffs are frequent – this region arose independently as a new industrial region.

As a microcosm of the technological changes that have occurred in much of China's economy since the early 1980s, Zhong'guancun's story provides valuable insights. Its experiences are noteworthy in two respects. First, with the arrival of MNCs in Zhong'guancun, the region has received 'free-rider' benefits through these firms' international network connections. Along with this result of the *globalization* of economic activity, increased *localization* of economic and social processes has been developed simultaneously. Second, the region could be appropriately described as a *new style administrative district* compared with the ones under the planned economy. It has not yet, however, turned into a 'technology-oriented complex' characterized by a process of synergistic interaction between new firms.

In the former Chinese command economy, the development model of the Zhong'guancun region was extremely hierarchical. Universities and research institutes were authorized by more than 40 supervisory agencies of different sectors both at the national and the local level. By the end of the 1970s, although there was great innovation potential in the area, bureaucratic, hierarchical organizations were seen to have inhibited innovation. Activities were redundant: similar research projects were usually repeatedly undertaken by different individual research groups, and there was hardly any interaction between them. Research results often remained in the laboratories, and only few were commercialised.

When reforms aimed at replacing the old economic system were introduced, they propelled a shift in organization from hierarchy toward networks. Zhong'guancun was quick to adjust, and appeared to be evolving into a genuinely entrepreneurial new-tech agglomeration. A noticeable innovative atmosphere and a bottom-up experiment emerged in the early

1980s. A set of reform measures and some market-oriented mechanisms were introduced in universities, research institutes, and state-owned industrial firms. In 1979, members of the Chinese Academy of Sciences (CAS) traveled to the USA and observed personally the success of Silicon Valley. Almost immediately, in 1980, a few scientists acted as risk-takers and established non-state-owned firms in the region. However, public opinion and social pressure still dictated that engineers and scientists should do what they were supposed to do – the planned research duties allocated by the research institutions or universities they belonged to; entrepreneurship had nothing to do with those duties. This situation did not change until the beginning of 1983, when the local government referred to the first Chinese non-state-owned firm, founded by a member of the CAS, as a model in a document guiding the further development of the Zhong'guancun region. Since then, new firms started by academic researchers have emerged one after another. By the end of 1983, 11 firms were operating in Zhong'guancun. The number of firms grew to 40 by the end of 1984, when the *Resolution for the Reform of the Economic System* was promulgated by the central government and attempted to strengthen the links between research and production in China (Wang and Wang, 1998).

Subsequently, universities, research institutes attached to the Chinese Academy of Sciences or ministries, and army- and state-owned large and medium-sized firms have increasingly set up technology-based and market-driven firms as 'try-out' business pioneers in order to convert their research results into products for sale. The entrepreneurial professionals make full use of four '*self-principles*' encouraged by the government – self-chosen partners, self-financing, self-operation, and self-responsibility for all losses caused by the venture – and actively set up joint ventures between non-state-owned and high- and new-tech firms with local governments and other organizations. The number of new-tech firms grew dramatically. In 1988 the Beijing New-Tech Experimental Zone (BEZ) was established within the Zhong'guancun region. The BEZ was zoned as a well-defined area of approximately 100 square kilometers by the Beijing Municipal Government. Because most powerful high- and new-tech firms of BEZ are concentrated in Zhong'guancun area, the distinctions between the administrative 'BEZ' and the address of 'Zhong'guancun' are usually blurred, and the firms are usually called Zhong'guancun's firms. As the first of 53 high- and new-tech industrial development zones approved by the State, its development has been stimulated since its creation by the new national policy aiming at industrialization and commercialization of research results (Wang and Wang, 1998).

As the most important knowledge-intensive region in China, the BEZ drew the attention of foreign MNCs already in its early stages. It has attracted such firms as IBM, AT&T, Intel, HP, DEC, Microsoft, General Electric, and Bell South of the United States; and Mitsui, Mitsubishi, Matsushita, Hitachi, NEC, and Canon of Japan. The number of foreign firms has increased rapidly (Table 8.1). Firms involving foreign capital have grown faster than other firms in this region. The total foreign capital invested in the BEZ has reached over US$ 1 billion, including registered funds of over $700 million, and direct foreign investments has reached US$ 400 million. The multinationals have exerted a great impulse on Chinese firms in the BEZ, speeding up their globalization process. With the penetration of multinationals, firms in BEZ are obtaining venture capital in joint research and development projects, improving their management style, and boosting the performance of qualified personnel. All these advantageous factors are prerequisites of innovation.

Table 8.1   Foreign direct investment in BEZ firms

Year	1989	1990	1991	1992	1993	1994	1995
Number of firms	36	70	140	300	700	780	806
Share of all BEZ firms (percent)	4.2	7.2	10.4	12.3	18.6	18.4	18.2

Source: Information Statistics Center of BEZ (1996, p. 13).

The average annual growth rate of the high- and new-tech firms recognized by the Management Commission of BEZ is 35.6 percent since 1988 (Figure 8.1). At the end of 1996 the total number of the high- and new-tech firms reached 4,506 and the annual gross income of the BEZ reached 25.8 billion yuan.[2] The total industrial output value of the BEZ accounts for 80 percent of the Haidian Administrative District's total and 8 percent of Beijing Municipality's total. The top 20 firms in the BEZ each attained a value of industrial output exceeding 1 million yuan (about $US 8 million). Domestic creative firms include Stone Group, Peking University Founder Group, Legend Computer Group, Beijing Huasun Computer Co., Beijing YaDu Sci-Tech General Co., Hope Computer Co. Ltd., Beijing Four-Ring Pharmaceutical Factory, Time Group, Tsinghua Unisplendour Group, and New Auto Group.

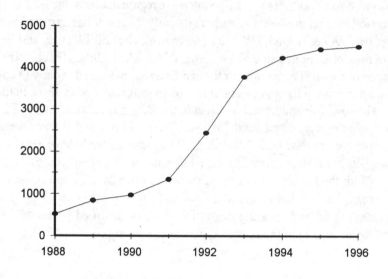

Figure 8.1   **Growth of high-tech and new-tech firms in the BEZ (1988-1996)**

Source: Information Statistics Center of BEZ (1996, p. 13).

Most are engaged in the electronic field. The industrial output value of the two leading sectors - information industries and photo-mechanical-electronic integration - account for 50 percent and 22 percent respectively of the industrial structure of the BEZ. This indicates that the BEZ has become a region specialized in electronics.

A comparison of the dimensions of bureaucratic, hierarchical organizations and network organizations that have potential in the Chinese context is shown in Table 8.2.

Figure 8.2 shows a Chinese hierarchy model in which the three sizes of circles stand for three levels of organizations. Shown by the solid one-way arrows in the figure, the higher-level organization controls the activities of the lower level; the lower-level organizations obey their authorities. There is almost no direct linkage between organizations at the same level. In reality, this pyramid of controlling relationships is more complex. Each organization

Table 8.2   Characteristics of the large ('old style') hierarchical state organizations and the ideal new private network organizations in the context of Chinese innovation

	Characteristics of hierarchical organisations	Ideal characteristics of network organisations
Relationship between higher and lower administrative levels	Many levels. Control and command relationships	Few levels. Appropriate or desirable to deal with lower levels; free to accept or reject offers from higher levels
Management style	Autocratic	Particpatory
Boundaries between actors	Fixed. Vertical integration	Permeable. Outsourcing and alliance
Forces driving innovation	Government command from above	Intense commercial competition from below
Financing	From government	From government and firms
Innovation process	Linear and sequential	Parallel and overlapping
Features of innovation	Inertia, low productivity, low quality	Speed, high productivity, high quality
Type of innovation	Technological	Technological and institutional
Benefit	Government	Owners, shareholders
Economic system	Command	Market

Source: Chen (1997, p. 30).

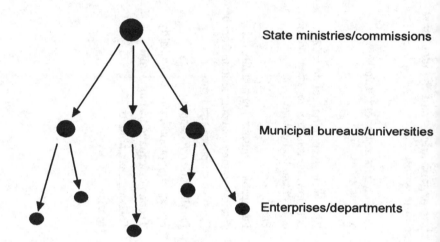

Figure 8.2   Hierarchy model in the Chinese context

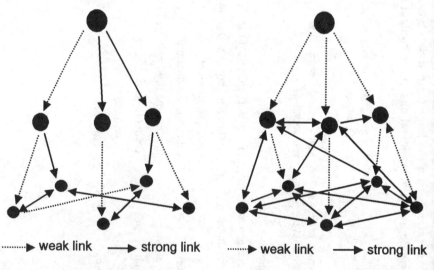

Figure 8.3   Weak local innovation network model in the Chinese context

Figure 8.4   Strong local innovation network model in the Chinese context

at a lower level could be controlled by *many* authorities at higher levels. This situation happens frequently in China, especially in Beijing where government bureaucracies at so many different levels exist. As the Chinese saying suggests, 'they have too many mothers-in-law'.

Figure 8.3 and Figure 8.4 show the potential transformation process from weak to strong local innovation networks in the Chinese context. Shown by the dotted lines in the two figures, leading organizations at all levels still administer, but control over their lower levels is relinquished to some extent, so that actors at lower levels are permitted to handle their own affairs more independently. As a result, shown by the solid two-way arrow in Figure 8.4, wide connections and co-operative relationships could be established among the actors with common goals. In other words, horizontal economic co-operation has grown stronger and closer network relations have gradually formed.

The process of new firm formation in the Zhong'guancun region can be recognized as a process of networking – reorganizing the resources of technological personnel and restructuring the system of science and technology (Wang and Wang, 1998). The following points are the main reflections of networking:

- Self-chosen partners: Most new-tech firms of Zhong'guancun comply with the four 'self-principles' encouraged by the government. In this case, people organized new firms with partners across the borders of their old administrative units.
- Several non-profit organizations, such as the Beijing New-tech Firms Association, and the Non-state-owed Enterprise Association, attempt to hold members' meetings regularly to exchange experiences about each other's operations. These organizations have kept their informality. These face-to-face circles operate to solve concrete problems of small firms: adapting to the changing policy in economic reform, seeking credit guarantees, or finding ways to deal with new situations. The informal information exchange seems to be increasing somewhat the chance of cooperation for innovation. However, many firms have spun-off from different state-owned tertiary institutions and maintain strong relations with their 'parents' under different government ministries or equivalents.
- Zhong'guancun's local innovation network, although at present still weak, could develop into a stronger one. The actors involved are small and medium-sized firms, universities and research institutes, and

government agencies, including agencies of industrial bureaus of central government and agencies of the Beijing Municipal government and the BEZ Management Commission. Resources related to innovation, such as capital, technology, information and personnel, should flow among all of these.

The important tasks in building the innovation network of the Zhong'guancun region are: to reduce hierarchical relations, to strengthen inter-sectoral links, and to continually develop co-operation between actors. The development of these links will tap the innovative potential of this region and facilitate synergy and cross-fertilization. To some degree these developments are already in progress.

Weaknesses in Zhong'guancun's Innovation Environment

Despite the increasing understanding concerning the non-linear nature of innovation processes (e.g., Asheim and Cooke, Chapter 6, this volume), there is still a misconception among Chinese policy makers that innovation follows a linear sequence. Firms have failed, therefore, to get the needed support for improving their capabilities in creating competitiveness. This factor clearly underlies Zhong'guancun's limited innovation capabilities. The suggestion that it should search for more innovative capabilities through the development of its local innovation networks has begun to capture the attention of the Beijing municipal government. There are impediments to such developments however. In the following, the recent policy bias toward hierarchies and large firms is seen as one such impediment. The discussion is followed by a characterisation of the existing (weak) local networks.

Bias toward Hierarchies and Large Firms

The large Chinese state-owned sector has faced a range of problems:

> ... the Chinese economic system has faced a more general choice between integrating activities within large enterprises and fragmenting them across distinct firms. This choice is quite different from the 'markets versus hierarchies' choice in market economies ... In China, the alternative to integration is not the market but rather planning and administrative allocation of flows of goods between separate entities. Second, the choice is not a natural one, although there may be empirical regularities in decisionmaking processes

and outcomes. It is subject to administrative fiat, and outcomes are influenced by bureaucratic conflicts and by administrative power, including sometimes the power of large firms. For these reasons, inefficient outcomes are only to be expected (Byrd, 1992, p. 329).

A current reform policy promoting large firms and invigorating small firms (*Zhua da fang xiao*) is aimed among other things at eliminating some of these obvious problems. An official statement explains the current policy:

> We will implement the large company and large group strategy, with the greatest efficiency as the principle, and brand-name and fine-quality products and enterprise groups as the carriers. We encourage association between strong enterprises and merger of inferior enterprises with superior ones, so as to concentrate resource elements on superior enterprises in the regrouping process (Wang, Z, 1997, p. 22).

This implies that government policy is currently aimed at supporting large firms. Such a policy unavoidably influences other parts of the economy, such as township firms (*Beijing Review*, 1997) and smaller non-state-owned high- and new-tech firms.

Having faced a decade of fierce rivalry between scattered organizations in the beginning of the economic reform during the 1980s, there is a shift to vertically-integrated corporations. At the outset, this seems to follow the path of the past experiences of merger and amalgamation in the west that has been fiercely criticized by Porter in the Western context. Porter's well-known view concerning a key factor in creating the competitive advantage of nations is the following:

> Conventional wisdom argues that domestic competition is wasteful: it leads to duplication of effort and prevents companies from achieving economies of scale. The 'right solution' is to embrace one or two national champion companies with the scale and strength to tackle foreign competitors, and to guarantee them the necessary resources, with the government's blessing. In fact, however, most national companies are uncompetitive, although heavily subsidized and protected by their government ... Domestic rivalries, like any rivalry, create pressure on companies to innovate and improve (Porter, 1990, p. 82).

Several years after Porter's observations and after ten years of promoting small-firm development, Chinese policy-makers have gone back to adhering to the approach of organizing large-scale, vertically-integrated. They seem unaware of Porter's criticism and of the Western experience that 'the key problem for small firms appears not to be that of being small, but of being isolated' (Pyke and Sengenberger, 1990, p. 4). The Chinese industrial system for the most part still remains in the late 1990's in the stage of mass production and mass consumption based on the exploitation of economies of scale. Until very recent times, one could frequently hear opinions such as: 'A system of small-scale firms in our country cannot compete with international giants in a global economy', 'Organize Chinese large corporations!' and 'Upsize!' Similarly, it is believed in China that only through vertical integration can small firms acquire the attributes possessed by larger organizations and overcome their deficiencies in capital and technology. Such a belief has resulted in an increased, rather than decreased, vertical integration and *a new style of hierarchical control* in the past decade. It is believed that favoring vertical integration over reliance on network relationships has a certain rationality in China. Several reasons are offered for this.

A Need for Vertical Integration   Vertical integration is a response to current conditions of Chinese economic and technological development. In other words, the former 'right solution' of one or two large firms that was criticised by Porter in the West is now seen as a necessary 'right solution' for China. It is believed that the principal result of the bitter competition between small high- and new-tech firms is that China can report few economies of scale and only poor synergy among its R&D and manufacturing activities.

A Need for Venture Capital   The development of most high- and new-tech firms has been sharply hindered both by a lack of venture capital and by difficulties in obtaining financial support from the government. The major firms in the BEZ region, including Legend, Founder, Jinghai, Daheng, Hope, and Stone, have had similar development paths. Their entrepreneurs, ambitious for their own innovative products, had to become involved at the beginning in commercial activities, such as selling computers and computer components and developing real estate, all for the purpose of accumulating capital. Both venture capital and loans are acquired only by those which have already become strong (i.e. already vertically integrated). In fact, those firms are able to get financing from the international financial market.

As Table 8.3 shows, 65 percent of the firms in the BEZ borrow from financial institutions. Since 1984, national financial institutions, such as the Industrial and Commercial Bank, Agriculture Bank, China Bank, China Construction Bank and China Insurance Company, have set up their branches

Table 8.3   Sources of finance for BEZ firms

Source of finance	Percentage of BEZ firms using this as principal source of finance
Loans from financial institutions	65.0
Foreign capital	12.7
Equity financing	7.6
Venture capital	5.6
Joint venture	4.1
Others	5.0

Source: Information Statistics Center of BEZ (1996, p.10).

in the Zhong'guancun region. By the end of 1993, all kinds of financial institutions had granted a total 3.8 billion yuan in loans to BEZ firms, including liquid capital loans of 3.18 billion yuan, and scientific development loans of 0.62 billion yuan. This is 2.8 times the loan level in 1988. By 1995, there were almost 200 financial organizations with about 4000 staff in the BEZ.

Need to Shelter Weaker Firms   Larger high- and new-tech firms could act as incubators and seed-beds for smaller ones. Within the BEZ, 88 percent of firms have a total income of less than 5 million yuan, indicating a strong small-firm-oriented structure (Information Statistics Center, 1994). Since the surrounding environment had changed so much in a short period of time, from a very rigid planned system to a partial and underdeveloped market, impulsive actions and lack of marketing experience were observed everywhere. The small firms are in a difficult condition, described in a Chinese phrase *jiu si yi sheng* (meaning 'among every ten firms only one firm survived').

In early 1990 Peking University had 15 high- and new-tech firms operating within it. Spin-offs started frequently, mainly in the applied science departments (or colleges) (Wang and Wang 1998).[3] However, at present, those firms have been concentrated into five groups (Founder, Ziyuan, Weiming, Qingniao, Weixin) which constitute the main body of the industry of Peking University. The experience of the Founder Group Corporation, a noted large and new printing corporation owned by Peking University, provides a good example of the current trend. In 1986, Professor Wang Xuan founded the Beijing Founder Electronic Co. Ltd. with the help of Peking University. A decade later, the corporation earned an annual net profit exceeding 130 million yuan. Its business scope has expanded into many fields of electronic publishing. The Group consists of 37 subsidiaries and a large-scale chain of over 300 distributors across the nation (Jiang, 1997). In 1996 it employed over 2,000 people, or about two-thirds of the 3,000 employed by Peking University's industry, and its industrial output value accounted for more than 90 percent of the University's total in 1996. Its performance can be attributed to the processes of acquisition and merger which have been encouraged by the University as well as by the central government. Founder is taking over firms both inside and outside of Peking University, leading to an oligopolistic structure. An executive of the group describes the situation: 'As it is more comfortable to have a rest in the shadow of a big tree, most small firms should stay under the shadow of a big firm'; or 'A car runs faster when it is on the highway'. Similarly, the Chinese Academy of Sciences' Tsinghua University also implements a large business strategy (Wang and Wang, 1998). It appears that through the merger process the number of firms in the BEZ will decrease in the future.

Such a trend towards more integration involves potential problems. Namely, it is possible that the trend towards mergers turns the Zhong'guancun region into a more hierarchical economy. Vertical integration could prove successful if large hierarchical firms also could increasingly give way to more decentralized organizations. In such organizations the transfer of intermediate responsibility to lower levels and its network would take place, and would be facilitated by new technologies. A staff member of a Peking University company which is to be merged said that he hopes his company could have some independence and flexibility after the merger.

As the merger process takes place, the economy of the BEZ is increasingly dominated by a small number of larger firms. The BEZ's four largest firms produce about 50 percent of the region's industrial output value. Figure 8.5 shows that the number of new registrations of high- and new-tech firms peaked shortly after the former Chinese leader Deng Xiaoping paid his widely recognised visit to South China in early 1992, indicating the overwhelming governmental enthusiasm to boost economic development by creating high- and new-tech firms, among other things. However, since then the number of startups has declined steadily. By contrast, the number of deregistered firms has increased quickly. Currently people involved in starting up new firms in the BEZ are difficult to find, indicating a lack of entrepreneurship. This suggests that the number of high- and new-tech firms could decrease continually in the future.

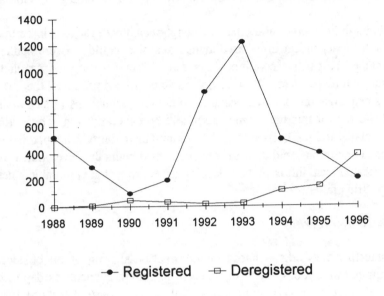

Figure 8.5   Number of registered and deregistered firms in the BEZ (1988-1996)

Source: Information Statistics Center of BEZ (1997).

The merger process discussed above is not proceeding without problems. Organizational differences cannot easily be broken down simply by integration, and it is necessary to take into account some of the costs of integration that are usually overlooked. Moreover, some 'small giants' that had performed well before mergers complain that name changes forced on them by higher hierarchical levels have joined them with less successful firms which adversely affects their previously good reputation.

In addition, processes of diversification (both into related and into unrelated businesses) happen when merger activities occur. Almost all larger firms in Zhong'guancun are seeking to extend their activities into other fields. For example, Founder Group is penetrating the chemical industry, and Stone group operates in the food industry. Because most electronics firms that borrow from banks cannot afford their high interest rates by producing electronic products alone, they have to make more money in other fields. This trend may inhibit innovations in the fields of specific technologies, such as electronics.

Although the merger activities in Zhong'guancun are rational in certain respects, the independent firm-based industrial system could be overwhelmed by changing competitive conditions. As demand for a variety of products increasingly replaces demand for cheap and standardized products, the large business organization that was designed to achieve economies of scale will present the risk of internal segmentation and bureaucratization. There is a danger that hierarchical corporate structures limit their ability to adapt quickly to changed conditions, and their inward focus also limits the development of a sophisticated local infrastructure, leaving the entire region vulnerable when the large firms falter.

Weak Local Networking

Local interfirm networks – based on information-sharing or on physical input-output transactions – can be detected in Zhong'guancun as depicted above. However, these constitute only a weak network pattern (Figure 8.3). Apart from price-report periodicals by certain information networks, information exchange in the Zhong'guancun region is facilitated through several non-profit organizations, such as the Beijing New-tech Enterprise Association, and the Non-state-owned Enterprise Association. Perhaps the most powerful organization is the Chinese Software Alliance, which has a role in protecting firms' property rights. However, the informal communications for information exchange do not seem to be increasing co-operation for innovation. An example of this point is Taisun Industrial Community, which was formed in 1994 by the representatives of 15 non-governmental firms, including famous 'small giants' such as Legend Group and Stone Group. However, the real purpose of Taisun Industrial Community is not to promote technological development. However, it has failed in its purpose to integrate finance, real estate and other businesses of the 15 firms concerned.

As for physical input-output transactions, there have been few direct inter-firm materials flows, such as supplying components or finished products to each other or sub-assembly. However, more and more big electronic accessories stores satisfy the needs of demanding customers of the locale and the rest of the country.

As illustrated in the preceding sections, however, as part of the process of new firm formation some forms of networking have been established. New flexible practices based on networking principles can be found, for example, in two companies located near the South Gate of Tsinghua University and hiring part-time personnel from local universities or research institutes. One is the Greatwall Enterprise Institute (GEI), a noted non-state-owned consulting company with a staff of less than ten, yet nearly 50 employees have worked or are working part-time in GEI for more than three years. Most of them are students or professors from Peking University, Tsinghua University, and Chinese People's University. At present, over ten postgraduates and three faculty members are working in the GEI, covering a range of specialties. Another example is Huafu Electric Power Corporation, established by a professional who formerly worked in a large-scale state-owned plant. Its staff is engaged in R&D activities and consists predominantly of students from and postgraduates working at Tsinghua University.

A few joint training projects also can be detected. For example, Stone Group, the largest non-state-owned high-technology firm in China, invested a large amount of money in 1994 to cooperate with Peking University to conduct two-year MBA training courses specifically for Stone higher-level staff. Another example is New Auto Group, which cooperated with Beijing Broadcasting College, Beijing Science University, Beijing Aerospace University, and others to jointly train postgraduates in high-technology fields by investing 1 million yuan in 1996 (Chen, 1997).

The links between universities and industry, particularly small business, have not been strong, however. The departments and schools of universities in the region generally are not as sensitive to local economic circumstances. Many graduate students have been interested in short-term commercial business, rather than other fields which would lead to the upgrading of Chinese technologies in the long term.

Co-operation between different actors in this region is still inadequate. Lack of co-operation has reached the point where a number of firms have established their own internal supply systems and services for their personnel. Scientific interactions between scientists and professors from different institutes, between different departments of the same institutes, or even between different persons with similar interests, are still rare. This isolationist situation, which steadily erodes the local innovation system, is becoming increasingly serious. In state-owned universities and research institutes, distinguished researchers working on a specific line of research receive a small amount of funding from the central or municipal government. The funds cover not only R&D expenditure, but also complement their salary. Researchers are only expected to report their research results to the foundations, or to publish in academic journals. Therefore, they prefer to do research projects in isolation rather than in collaboration. Competition for such government funding is increasingly intense, causing a serious waste of capital and human resources. The weak local linkage formation in this region is caused by two different factors. First, local firms source production factors mainly from outside of Beijing rather than from the local region. Second, the branches of multinational firms are based outside China – in the USA, Germany, Japan, Singapore, Denmark, Australia, and Taiwan (Wang and Wang 1998).

There have been few successful cases of co-operation between the high- and new-tech firms of Zhong'guancun and those state-owned firms run under the shadow of Beijing's old management system. The essential difficulty lies in their incompatible operational systems. The difficulties in finding collaborators in the local large firm segment have lead new- and high-tech firms in Zhong'guancun engage in networking relationships with more active partners in other regions. Many traditional state-owned firms in Beijing have failed to compete with firms from other provinces in forming links with promising Zhong'guancun firms. Two examples of this are Tsinghua University and Beijing Aerospace University. In 1995 alone, Tsinghua University signed 124 contracts with the provinces of Hebei, Guangdong, Shandong, Zhejiang and Jiangsu for joint projects. Beijing Aerospace University has been networking with firms from the provinces of Guangxi and Shandong (Chen, 1997).

The high- and new-tech firms in Zhong'guancun tend to look outside Beijing for co-operation for reasons related to land requirements and industrial linkages. The office/shop rent rate and tax levels are higher in Beijing than in many other cities. There is both more room and greater flexibility to do business in other provinces, particularly those far away from the capital city. Moving out of Beijing helps the high- and new-tech firms get rid of some rigidity and formality of the old system that they otherwise would have to deal with in Beijing. It also enables them to take advantage of informal, and in some cases illegal, channels to source their production factors, even internationally.

As local networking among domestic actors in Beijing's new-tech agglomeration has been modest, firms are forming links with multinational firms for alternative support. China's 'small giants' in the field of high- and new-technology have already co-operated or collaborated with multinationals. Also almost all universities and research institutes in the Zhong'guancun region have established co-operative relationships with foreign corporations. By the end of 1995, there were more than 1,100 foreign investors in the BEZ, including 820 Sino-foreign joint ventures, and five Sino-foreign co-operative firms, of which nearly 800 were registered as featuring new-tech industries.

IBM has set up two research institutes in the Beijing ShangDi Information Industry Base, a new industrial park in the BEZ: the Chinese Research Center of IBM, and the DingXing Information System Development Co. Ltd., a joint venture with the Software Technology Center of Tsinghua University. The reason for IBM's location of an R&D center in the ShangDi Information Industry Base, close to Tsinghua University, lies in the region's technology-intensiveness (Chen, 1997).

Networking for Innovation: The Policy Perspective

This section looks at how the transformation of Zhong'guancun into a genuine innovation region could take place, based on the establishment of a strong local innovation network. Local actors in developing countries (as in developed countries) should become deeply embedded in the local economy and create a network of sophisticated, interdependent linkages, so as to generate self-sustaining growth of the local cluster as a whole (see also Fromhold-Eisebith, Chapter 9, *this volume*). Otherwise, local clusters such as Zhong'guancun would be merely a set of weak nodes within international networks of powerful multinationals. For developing countries like China, building local innovation network is necessary. The following are the main points set forth in a recent paper to the Beijing municipal government (Wang, 1997):

- The universities and research institutes of Zhong'guancun have been isolated from the local economy for several decades. Since the early 1980s their resources have been opened to society to some extent; however, many of their research results have disseminated outside of Beijing to other cities. Even local high- and new-tech firms have not fully used the local resources. In other words, Zhong'guancun is still an enclave outside Beijing's economic system. In order to develop Beijing's local economy as well as China's economy, the actors in Zhong'guancun must recognize the significance of local networking.
- There is still a host of administrative restrictions limiting the exercise of dynamic entrepreneurship. Because entrepreneurial development of this region heavily depends on its sustainable innovative atmosphere, support by local government for entrepreneurial activity and innovation is necessary to nurture the development potential that exists in the Zhong'guancun region.

- There is a need to build a network of mediating actors to operate between vertical integration and spot market transactions, with the aim to provide a more stable basis for co-operation. These actors would provide services of financing, technology transfer, accounting, and consulting. In Chinese circumstances this could not be achieved without government support.
- The behaviors of individual actors should be bound by common rules related to networking. It is recognized that there are two aspects of such rules: the formal (new regulations) and the informal (business ethics and civic spirit).
- Universities and research institutes should open their doors more widely to the outside world. They should share their libraries, laboratories, engineering research centers as well as human resources with firms (especially small firms) located nearby. This may require further reforms within both the educational and the science and technology systems.
- Sites for face-to-face interactions also would be beneficial (Crevoisier, Chapter 2, *this volume*). There are many potential interaction sites in the Zhong'guancun region, such as meeting rooms, exhibition centers, trade markets, electronic accessories stores, and restaurants, but these are not used fully. In addition, the various associations of firms should be supported by local government.

To sum up, the government's intention should be to create national and local synergy effects. In practice, administrative ties between authorities and firms still hinder independent decision-making by firms and maintain dependent relationships. Therefore, the transformation of the administrative commission of the BEZ into a real service institution would be beneficial for the local economy.

Conclusion

This chapter has focused on the possibilities of networking in creating innovative capabilities especially in the Zhong'guancun region. It is suggested that viable local innovation networks should develop in China as it moves toward a more market-oriented society. It seems premature to suggest that a genuine innovative environment exists in Zhong'guancun or elsewhere in China. Based on the discussion, several possible reasons for the lack of such an environment can be suggested.

First, there is an important difference between the deregulation process occurring in Western market economies and in China. The deregulation typically seen in an old market economy is a process whereby some parts of a regulatory framework with supposed negative impacts on efficiency are altered or removed to release the vitality of the economy. In China, when the rigid planned system was broken down by a deregulation process and the vitality of production was released, there was no pre-existing body of laws and regulations suitable for the new market system concerning property rights, commercial and financial laws, and share-holding (see also Barta, Chapter 7, *this volume*). This was particularly true in the Zhong'guancun region where both the products and the kinds of producers were new. As a result, entrepreneurs stood in a brutal developing market environment, struggling against unfavorable influences from both the old and new systems. In such circumstances, and facing market competition, often the most important entrepreneurial struggle is not to innovate, but to survive. As a result, the lack of laws and regulations suitable for a market economy, together with unclear and inconsistent government policy, have hampered local networking for innovation among firms in Zhong'guancun.

Second, as far as co-operation is concerned, the changing ownership relations in domestic firms are worthy of notice. A large number of high- and new-tech firms in Zhong'guancun have spun off from their state-owned parents and have been receiving strong support from their parent organisations, or from universities. Such strong vertical ties work against the formation of networking among new innovative firms. These firms are only 'semi-detached' from their parents, and become easily enmeshed back into the old hierarchical systems.

Third, one of the roots of the problem lies in a general lack of understanding in China's financial community concerning the need to put money behind technology, and to show patience in waiting for the investment to pay off. This has caused a vicious circle among firms in this region. The personnel of firms, universities and research institutes spend a great deal of time looking for money from trade, real estate development, or high-interest loans from the banks, but they have few initiatives in innovation.

Fourth, although Zhong'guancun is preparing to enter the information age, the local culture seems behind the era. It is evident from Western experience that the ability to synthesize complex technologies seems clearly to require trust-based relationships among actors. Familial ties are not necessarily enough to nurture the needed trust-based relationships in Asian economies as has been suggested in the literature (Fukuyama, 1995). Other trust-based relationships are not easy to establish, however, for several reasons. The historical tradition of federal separation caused a divisive culture in the country. Also, past divisive political campaigns, such as the Cultural Revolution, seriously damaged the inclination to develop trust-based relationships between individuals. In addition, the conservative ideas of Beijing officials have a negative impact on the local culture.

In order to generate innovation and self-sustaining growth in the local economy the local actors of the Zhong'guancun region need to create various forms of local network relations. Building innovation networks in the local economies of developing countries like China is difficult but it is a necessity. Otherwise, local economies are reduced to sets of weak nodes in the webs of powerful multinationals.

Notes

1   Two types of 'development zones' were initiated by two different authorities. One is the Economic and Technological Development Zones (ETDZs) initiated by the Administrative Office of Special Economic Zones attached to the State Council. The other is the High- and New-tech Industrial Development Zones (HIDZs) initiated by the Science and Technology Commission. Originally, the former was planned as production-oriented, meeting the needs for economic growth through attraction of foreign investment. The latter were planned as science parks where Chinese technological personnel could take an important role, bridging the gap between national technology and production.

2   The annual gross income indicates the annual gross income of technology, industry and trade. It is defined as the sum of annual sales income from firms' three activities: manufacturing, technology transactions, and other trade.

3   Here, the term spinoff is used in a broad sense. According to Glasmeier (1987) and Roberts (1972), spinoffs are new ventures started by entrepreneurs who previously worked at another high-tech organization; many product ideas are developed while working for the parent and become the base of a spinoff when the parent decides to abandon them (Roberts 1972). In the Chinese context, 'another high-tech organization' may include universities and research institutes.

References

Beijing Review (1997), 'Township Firms Growth in Size', *Beijing Review*, vol. 40, no. 3, pp. 20-21.

Braczyk, H.-J. and Heidenreich, M. (1998), 'Regional Governance Structures in a Global World', in P. Cooke, H.-J. Braczyk and M. Heidenreich (eds), *Regional Innovation Systems*, London, UCL Press, pp. 414-440.

Byrd, W.A. (1992), 'The Anshan Iron and Steel Company', in W.A. Byrd (ed.), *Chinese Industrial Firms under Reform*, Oxford University Press, New York, pp. 303-370.

Chen, Y. (1997), 'The Role of the University in Regional Innovation Network of Zhong'guancun', Master's thesis, Research Institute of Science and Society, Tsinghua University, Beijing, 100831, P. R. China.

Fukuyama F. (1995), *Trust: The Social Virtues and the Creation of Prosperity*, Free Press, New York.

Glasmeier, A. (1987), 'Factors Governing the Development of High Tech Industry Agglomerations: A Tale of Three Cities', *Regional Studies*, vol. 22, pp. 287-301.

Information Statistics Center of BEZ (various years), *Research Report of BEZ*, Beijing, Information Statistics Center of BEZ.

Jiang, W. (1997), 'Liberating Chinese Printing with Keyboard', *Beijing Review*, vol. 40, no. 19, pp. 13-16.

Porter, M. (1990), 'The Competitive Advantage of Nations', *Harvard Business Review*, vol. 68, no. 2, pp. 73-93.

Pyke, F., Becattini, G., and Sengenberger, W. (eds.) (1990), *Industrial Districts and Inter-firm Co-operation in Italy*, International Labour Office, Geneva.

Qian, S. (1991), 'Development of China's Nuclear Science', *Beijing Review*, vol 34, nos. 7-8 (Feb. 18-March 3), pp. 34-40.

Roberts, E. (1972): 'Influence on the Performers of New Technical Enterprises', in A. Cooper and J. Komives (eds), *Technical Entrepreneurship: A Symposium*, Milwaukee, WI, Center for Venture Management, pp. 126-149.

Si, N. (1997), 'Technology Emphasized', *China Daily*, May 21, 1997.

Sui, Y. (1992), 'Zhongguo Keji Jingji Fazhan ji Yunxing Jizhi de Goujian yu Wanshan' (Establishment and Improvement of Coordinate Development and Operational Mechanics of S&T Economy), *Keji Daobao (Science and Technology Review)*, no. 7, pp. 44-47.

UNIDO, (1996), *Industrial Development: Global Report 1996, Executive Summary*, Oxford University Press, New York.

Wang, H. (1991), 'China's Industry: 42 Years Versus 109 Years', *Beijing Review*, vol. 34, no. 39.

Wang Jici and Wang Jixian (1998), 'An Analysis of New-tech Agglomeration in Beijing: A New Industrial District in the Making?' *Environment and Planning A*, vol. 30, pp. 681-701.

Wang, J. (1997), 'Guanyu Zhong'guancun xin jishu quyu fazhan wenti de shenceng sikao' (Policy thinking on the development of Beijing's Zhong'guancun new-tech region), in Jing Tihua (ed), *Beijing jingji xingshi fenxi yu yuc: 1997 lanpishu (Analyzing and Forecasting Beijing's Economic Situation: 1997 Blue Book)*, Capital Normal University Press, Beijing, pp. 106-125.

Wang, Z. (1997), 'State Firm Reform in a Major Stage', *Beijing Review*, vol. 40, no. 9, pp. 19-22.

9 Bangalore: A Network Model for Innovation-Oriented Regional Development in NICs?

Martina Fromhold-Eisebith

Introduction

In the face of a growing variety and variability of successful innovative regions worldwide, it is increasingly difficult to claim that there is a single path of regional technology-based industrial growth. This situation is further complicated by the fact that in addition to the long-known success stories of regions in industrial countries, a number of technologically competitive areas also are emerging in developing economies. Among these is the region of Bangalore in the southern Indian state of Karnataka, the home of a growing cluster of companies in information technology (IT) sectors and celebrated in the media as a new 'Silicon Valley'. The Bangalore case provides a prominent example challenging further investigation.

There is no doubt that the specific pattern of another 'Silicon Valley' cannot be repeated elsewhere in the world, especially not in a different place, time and culture. But are there general 'typical' characteristics that could be regarded as broadly applicable to innovation-oriented regions in the industrialized world? And to what extent are they also valid and traceable for developing countries? This chapter tries to shed some light on this issue from the theoretical as well as the empirical side, searching for the crossroad between an individual Indian case study and a general model for the developing world.

In order to do so, we focus on specific core aspects of successful technology-based regional development, based on the extensive research on factors promoting the spatial clustering of innovative firms in industrial countries. In recent years special emphasis has been placed on the relevance of regional networks and systems of collaboration as promoters of continuing success. Networks of a specific, information-intensive kind are now widely considered to be a crucial characteristic of innovative regions, and to provide the necessary local embedding of globally competitive economic activity in the post-Fordist era of flexible specialization (Conti, 1993, 1997; Oinas and Malecki, Chapter 1, *this volume*).

Though the general applicability of networks still remains to be empirically proven even for the industrial world (not to mention the methodological traps to be circumvented to capture the relevant aspects), the transposition is tried: How valid is the networking hypothesis for technology regions in countries which are just recently experiencing the thrust of modern industrialization, such as the Newly Industrializing Countries (NICs)? It appears that, especially regarding local linkages, the ideal models do not fit NIC regions, and conceptual and theoretical development seem necessary (Park and Markusen, 1995; Park, 1996). Moreover, in the context of less developed countries, regional linkages and collaboration are all the more important as these economies mostly need a 'trickling down' and diffusion of technological impulses via structures of an industrial district type to proceed in their development process (Schmitz and Musyck, 1994). Knowledge-intensive collaborations between firms enhance the manifold learning processes that are especially urgent for a developing country, also compensating for the missing support from other (e.g. administrative) actors in place.

This chapter is divided into two parts: The first looks at theory, and tries to derive basic traits of network formation in technology regions of NICs. The general chances for regional information-intensive contacts are evaluated against the background of general developmental strategies and trajectories characterizing (Asian) NICs. These shape their 'national innovation systems' (Nelson and Rosenberg, 1993) and are reflected in other processes, such as network behaviour and regional development, as well. We could then expect regional networking features to be significantly distinct from those in industrial economies. To fill some flesh into the theoretical frame, the second part presents the case study of the Bangalore region, integrated into the Indian national innovation system. The system of networks in the technology region of Bangalore is described based on an empirical investigation of the contacting behaviour of important local actors. This case study shows the particular expression of general conditions of technology-oriented industrialization. This research not only demonstrates that there is some significant embedding of firms taking place in a developing country, but it also contributes to a deeper understanding of the very nature of 'embedded' relationships, a phenomenon still needing further exploration (Oinas, 1997).

Theorising Network Formation in Technology Regions of NICs

Though there has been much research and discussion that attempts to derive models of successful innovation-oriented regional development, one field is widely neglected so far: that of concepts specifically relating to less developed countries. It has to be admitted, however, that the emergence of technology regions[1] in less industrialized countries is a quite recent phenomenon, accompanying the ambitious objective of mostly NICs to achieve 'systemic competitiveness'. This refers to economic growth based on high quality, innovative products manufactured with the help of new technologies and a highly skilled workforce, instead of standard products as an outcome of low-skilled mass production (Esser *et al.*, 1992). Only a few authors have carried out case studies on regional development in NICs using contemporary concepts such as flexible specialization or industrial districts (Nadvi, 1992, 1994; Park, 1996; Park and Markusen, 1995; Rasiah, 1994; Schmitz, 1990). Also, Bangalore and its engineering firms have been investigated in this context (Holmström, 1994).

A systematic approach that also provides a general framework for further empirical studies is still missing. The following sections sketch the general outlines of such a theory of network formation in technology regions of NICs.[2] There are two argumentative steps: First, the main characteristic of successful innovative regions in industrial economies is briefly summarized. Second, the possibilities of advanced developing countries to seize these opportunities are evaluated with respect to specific characteristics of their technology-oriented industrialization.

Knowledge-Intensive Networks: The Key to Successful Technology Regions in Industrialized Countries

The description of the main factors enabling or enforcing regional concentrations of industrial innovation in developed countries is only a necessary first step for translating 'Western' theories to less developed countries. Therefore, discussion of the wide range of different approaches and concepts, some of them still under fundamental discussion, is beyond the scope of this chapter (for an overview see Sternberg, 1995). But some common characteristics in these approaches can be identified. Of those, the following account concentrates on one central concept: network relationships.

Technology-based companies operate in an environment of growing information and learning requirements for achieving global competitiveness, especially if they are of the emerging vertically disintegrated, flexibly specialized kind (Foray and Freeman, 1993). Consequently, economists as well as economic geographers now put a strong emphasis on the importance of nearby sources of knowledge and support as universities, public/private research institutions, and industrial promotion organizations. By interacting with local firms they create synergies and self-reinforcing processes of innovation (Conti, 1997; Stöhr, 1986). On the one hand, 'endogenous regional development' (Brugger, 1990) is promoted via the transfer of technology from knowledge centres to local industry. On the other hand, such factors as availability of skilled people and knowledge from local institutions of education and science, high-level producer services, or the 'soft' factors of a good quality of life (Delaplace, 1993) attract exogenous investment by high-tech companies.

The embeddedness of firms in local systems of information exchange retains and even gains importance in spite of economic globalization, since networking opportunities are considered to be decisive for the 'pinning down' of globally-oriented companies in a specific area, and international competitiveness is established and secured by specific agglomeration economies (Amin and Thrift, 1994; Conti *et al.*, 1995; Maskell, Chapter 2, *this volume*). The existence and patterns of regional networks have therefore become the main focus of contemporary interest in the factors of success of innovative localities (Conti, 1993; Malecki, 1997; Oinas and Malecki, Chapter 1, *this volume*). It seems as if these relationships are important not only for a region to become, but especially to remain, innovative, since they encourage continuing learning processes of the resident companies in an evolutionary, self-sustaining way, combining knowledge external as well as internal to the region.

The notion of networks remains only partially understood. Two concepts offer insights into certain requirements to be fulfilled by the relevant linkages in order to promote technology-based success. In the form of an *industrial district*, the regional contact systems basically are comprised of companies in related sectors that exchange goods, services and information along the value chain (Park, 1996; Pyke and Sengenberger, 1992). The synergistic interaction of different complementary organizations on the local scale is captured under the label of the *creative or innovative milieu*, referring to the fabric of firms, universities, R&D institutes and administrative bodies (Camagni, 1991, 1995; Fromhold-Eisebith, 1995). There remains considerable uncertainty about the specific nature of the relevant linkages between local actors that makes them so peculiarly effective. But the following decisive commonality is emphasized for both relational concepts: The linkages have to be both information-intensive and based on trust – built to a large extent by informal and personal relationships and face-to-face contacts. As a quality distinguishing 'true' (or Marshallian) districts and milieus from other kinds of local supplier-customer networks, the local actors are culturally bound together by common representations and interests as well as by social coherence and a sense of belonging – features deeply depending on spatial proximity (Camagni, 1991; Conti, 1997; Fromhold-Eisebith, 1995).[3]

This brief discussion aims to show that the central characteristic of knowledge-intensive networks is generally valid for most successful technology regions in the industrial countries.

Towards a Theory of Network Formation in Technology Regions of NICs

We now consider whether nations with a fundamentally different framework of technology-oriented industrialization – Asian NICs of an older or younger generation[4] – can also follow the path of sustainable technology-based regional development based on knowledge-intensive relationships between local firms and a set of innovation-encouraging institutions. The discussion is carried out in light of some key topics derived from development theory.

In most Asian NICs, the precondition of regional clusters of support institutions favoring technology-based economic activities appears to be met. The validity of 'Western' regional theory has to be questioned, however, since the the emergence of local innovation networks seems to be confronted with severe constraints. Nevertheless there are quite a few examples of prospering technology-based industrial clusters.[5] This gives rise to the call for a new 'family' of regional theories that reflect the specific advantages and disadvantages of the configuration of innovation systems found in the NICs.

The relevant aspects of technology-oriented industrialization of NICs all contribute to their specific *national innovation systems*, the 'set of institutions whose interaction determine the innovative performance ... of national firms' (Nelson and Rosenberg, 1993, p. 4). These systems comprise organizations, their behaviour as well as policies shaping the creation and dissemination of industrial technologies in nation-specific ways. They have an impact on regional development. Each of the topics below (see Figure 9.1) relates to a

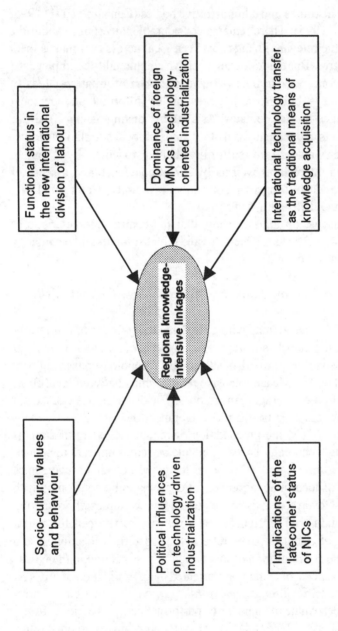

Figure 9.1 Factors influencing internal linkage formation in technology regions of NICs

range of theories depicting complex interdependencies among factors. In this chapter the impact on regional network formation is outlined.

Functional Status of NICs in the Process of Economic Globalization – The New International Division of Labour A first sphere of influence (not in the sense of a ranking) is represented by the 'new international division of labour' and the functions devoted to less developed countries accordingly (Fröbel *et al.*, 1986). With factor costs and factor availability as central variables, mostly low level economic activities get concentrated there, characterized by high labour intensity but low labour qualification. Additionally, the performance of local firms mostly has to be categorized as low-technology. Foreign multinational companies (MNCs), actually the engines of the functional separation between the developed and the developing world, often use sophisticated but mature production technology, with a prevailing dependency on external know-how (Dicken, 1992).

Local knowledge-intensive networking is obstructed for several reasons: The international separation of different phases of production leads to a mostly outward-oriented linkage behaviour of the firms, sometimes further reinforced by a strong export orientation. The standard production mainly concentrated in developing countries by MNCs usually requires much less external information-intensive contacting than leading and control functions such as headquarters, product design, and R&D. And there is a 'vicious circle' of continuing dependency of even advanced developing countries on external inputs of orders and know-how.

But the Asian NICs seem to be given a chance to follow the trajectory of industrial countries in increasing innovation-oriented activities of a flexibly specialized type, connected with a growing demand for high-level collaboration (Schamp, 1993). In many Asian countries, we already see clear signs of technological upgrading of local and foreign industrial activity, attracted by a growing supply of skilled labour (Dicken, 1992). Though the dispersion of R&D functions by MNCs in general lags considerably behind the externalization of production, some R&D investment is slowly sprouting in NICs, if the local conditions are favourable (Sachwald, 1995). This also gradually improves the chances for local knowledge-intensive linkages in these countries, especially when more and more 'home-base-augmenting units' of R&D get located in NICs, which have the explicit task to source region-specific knowledge for the mother company (Kuemmerle, 1997).

Dominance of Foreign MNCs in Technology-oriented Industrialization
Because of the overall dominance of foreign-based MNCs in the technology-based industrialization of developing countries (Buckley and Clegg, 1991; Marton, 1986), they are also the main actors in many of the NIC 'technology regions' identified so far. This means that national and regional development depends on actors whose ambitions are mostly oriented towards action in the global competitive environment. From their perspective, the branch location in a developing country is often regarded as only one element in an international company structure, searching to exploit location-specific advantages and externalities, and bound into worldwide supplier and customer relations (Dunning, 1993).

Under these conditions it has to be questioned if MNC branch plants are able or even willing to integrate locally beyond making use of the local labour market. In particular, the chances for subsidiary executives to join personal contact systems within their locality look quite slim because the management of MNC branches in less developed countries is so often comprised of expatriates from the MNC headquarters. The general behaviour of MNCs concerning dissemination of technological information to host developing countries is characterized by efforts to strictly control the outflow of knowledge and to keep the necessary know-how within the company boundaries (Dicken, 1992; Marton, 1986). Only if there already is a regional base of technological capacity do MNC branches show the propensity to link up to it and use – and to reinforce – the local accumulation of knowledge by R&D investment (Cantwell, 1993; Casson, 1991). Thus the likelihood that foreign MNCs integrate into regional networks of information exchange is limited to NICs which have already managed to build up a critical level of indigenous technological competence.

International Technology Transfer as the Traditional Means of Knowledge Acquisition   In developing countries, the acquisition of technical know-how for industry traditionally depends to a large extent on imports from industrial countries. Since the 1960s many developing nations have put some effort into increasing the indigenous base of know-how and technological capabilities (Baranson, 1969; Enos, 1991). But both the process of technology acquisition and the low level of indigenous competence have hindered improvements in technological capability.

MNCs also play the dominant role in technology transfer, for they have been and remain the main providers of state-of-the-art technology for most Asian NICs, via direct investment, licences, consulting, or sales of sophisticated equipment (Soesastro and Pangestu, 1990). The disadvantages of this dependency on international technology transfer have long been acknowledged by the developing countries as well as development organizations. Since a predominant concern was the inappropriateness of imported technology to developing country conditions, building up indigenous competence in R&D was seen as a political necessity to make better local use of the imports (Lall, 1992). Accordingly, the efforts undertaken were mostly directed towards adopting foreign technology and adapting it to local conditions and requirements, which often meant making it simple, easy to handle and labour-intensive. Therefore, the capacities related to appropriate technology were quite distant from both modern industrial technology requirements and the need for continual upgrading through knowledge-enriching collaborations with indigenous research institutions.[6]

On the other hand, there have been ambitious projects of building up state-of-the-art scientific activities in developing countries for a long time, sometimes with remarkable success. But the accumulation of scientific knowledge has been accompanied by a notorious and prevailing lack of national technology transfer to production (Silveira, 1985). This chronic gap between science and (private) industry widely remains to be bridged today. The scenery is changing, however. The Asian NICs in particular have succeeded in developing a considerable technological capability and industrial expertise and sophistication via imitation and reverse engineering. To stand international technological competition in export-based industries, they have established application-oriented R&D institutes which can now indigenously provide the relevant know-how transfer and integrate international technology transfer as complementary to indigenous sources for industrial progress (Bell and Pavitt, 1993; Chowdhuri and Islam, 1993).

Implications of the 'Late-comer' Status of NICs   Another factor possibly affecting regional linkage formation is the common characteristic of NICs to be late-industrializing compared to industrial countries. This implies a quite different global economic and technological setting of industrial development. Though it is to some extent connected with technological disadvantages (Bell and Pavitt, 1993), it also gives them some advantage in establishing up-to-date or even advanced competences in technology-based industries on a global scale, forcing them to further move the frontier of knowledge by their own R&D (and research collaborations). Thus the 'late-comer' status of NICs can even be interpreted as favouring joint efforts of technology actors on a national or regional scale.

The chances for less developed countries to 'catch up' emerge from the observation of long (Kondratieff) waves of economic development shaped by incremental and pervasive improvements of a basic technological innovation. Over time, the performance of actors within this 'techno-economic paradigm' increasingly approaches an ideal 'best practice' status (Perez, 1985). For late-developing countries it has been possible, for example, to enter the currently dominating paradigm of microelectronics at a high level of sophistication, since the barriers of entry are quite low. The necessary basic knowledge has been accumulated and published worldwide, and even recent product innovations can be bought at affordable costs (Perez and Soete, 1988).

Therefore the NICs that have only recently used their open technological 'window of opportunity' have been able to save time and expense compared to those industrial countries that had to develop the technology from scratch. In some fields ('hard' and 'soft' applications of microelectronics) some even managed to gain a lead over industrial countries, also by catching up in R&D.[7] Because of intense global competition and short product life cycles in this sphere, internationally transferred know-how is hardly available any more, forcing companies in NICs to develop technological advancements themselves. They have to draw the necessary knowledge from their own R&D and collaborations.

This argument again points to the necessary precondition of a threshold amount of technological knowledge already accumulated in a country and region. Only if there already is some initial capacity in terms of human skills and institutional infrastructure, a self-reinforcing collective learning process of regional actors can be stimulated.

Political and State Influences on Technology-driven Industrialization   In developing countries it is generally the state that takes over the responsibility to create skills and infrastructure as prerequisites for technology-based industrialization. Therefore the influences of policy on one hand and the activities of the public sector on the other have been particularly important factors in the successful take-off of NICs. Asian countries show a particularly strong bias towards state intervention and guidance, with a strong sense of pragmatism (Chowdhuri and Islam, 1993; Schmiegelow, 1990).

The state sets the political framework for development by a wide range of instruments regarding industrial and regional policy, science/ technology policy, and educational policy. Most of the NICs have strongly regimented economic activity, often following the objectives of import substitution in early phases and export promotion in later phases of industrialization. Accordingly, the nature and behaviour of foreign MNCs has been strongly restricted at first, followed by subsequent phases of liberalization and encouragement by incentives. Regarding MNCs, we can speak of a pronounced 'linkage-oriented policy', since here the main motivation has been to increase the positive techno-economic effects of MNCs via intensified linkages to indigenous firms (e.g. 'local content' requirements) (Marton, 1986). But the target level of such a policy has been national, not regional (though it has regional outcomes). Explicit regional policies have mostly been oriented towards the reduction of spatial disparities, devoting public financial resources to fulfilling the basic needs in backward areas instead of installing innovation-oriented policies in regions that already have technological potential.

An important basis for technology-driven industrialization could be created by NIC policies towards higher education and R&D in engineering. Here the Asian countries have been especially prominent in considerable improvements in infrastructure and output. But regarding regional impacts of this potential, they do little to promote endogenous regional development by the stimulation of local technology transfer linkages or spin-offs, as a diffusion-oriented policy would suggest. The mission-oriented approach prevails even with respect to 'technology regions', illustrated by the establishment of ambitious 'Technopolis' projects as important instruments of innovation policy. A whole cluster of sophisticated infrastructure is built virtually overnight, bearing only small chances for the meshing of local transfer networks over time.

In many NICs, the state plays an important role in technology-based production through public sector enterprises (PSEs). They were often expressly established as engines of technology-oriented industrialization, either in strategic sectors where private investment was excluded or in sectors where the initial costs were too high for private activity. Therefore their structure and behaviour regarding the use of new technologies usually contrasts to those of private firms in the country (James, 1989). PSEs, though frequently blamed for their ineffective operation, are especially noteworthy for having a long tradition of collaboration with foreign partners as well as with national R&D institutions. This makes them an ideal network intermediary, sourcing international knowledge which is then adopted and modified within the national/regional interaction system while enriching it.

Socio-cultural Values and Behaviour   Since industrial districts and creative milieus both are characterized by social linkages and socio-cultural coherence among their actors, this issue deserves special attention in discussing network formation in NICs as well. These aspects represent the important 'soft' factors being increasingly emphasized in accounts of regional success, though there is still considerable lack of knowledge about their detailed nature (Oinas and Malecki, Chapter 1, *this volume*). Since there is only a blurred picture of the causal role of culture in regional development in general, it is difficult to argue about its significance in less developed countries. Some claims can be put forward, however.

First, within and outside the Weberian tradition, attention has been paid to the interdependency of religion and economic development. A climax of this discussion probably was the 'discovery' of the decisive role of 'Confucian Capitalism' for the success of prominent Asian NICs, but this assumption has to be viewed with caution (Chowdhuri and Islam, 1993). Devotion to work and an ambition to learn can be understood as the usual concomitants and symptoms of a materialistic pursuit of money and status - quite basic human motivations regardless of religion. This could then explain why also countries that are not Confucian at all, such as India, have successfully built up innovative industries on the base of an ambitious, learning-oriented workforce. The attention should instead be shifted from religion to a more fundamental level of social and behavioral conventions (Storper and Salais, 1997).

Regarding the culture of networking behaviour, it can be suggested that, in less developed countries, personal contacts in general tend to be more important for business than in advanced industrial economies. This refers to the significance of face-to-face-communication as well as to emotional connotations of informal relationships like personal sympathy, confidence and trust, or the interest in a contact person's family and private, often spiritual world. When seen from this perspective, firms in NICs seem to have an even better starting position for networking than in industrial countries!

In conclusion, while the building of local knowledge-intensive linkages for NICs appears to be restricted, largely due to the dominant role of foreign firms and know-how, many NICs seem to be able to gradually improve in the course of industrial development by building up local high-level R&D institutions and capabilities. The line of reasoning provided above suggests that this is likely to happen through the emergence of specific patterns of regional networks. They reflect the national innovation systems that are specific to each NIC and distinct from those in the industrial countries with respect to the nature of actors involved, the kind and direction of knowledge flow within collaborative relations, and the significance of personal relations.

Bangalore as a New Generation Technology Region

The emerging technology region of Bangalore is chosen as an example to explore the validity of the hypothesis put forward above. India's fifth-largest city, with five million inhabitants, and the capital of the South Indian state of Karnataka, is by now well-known for its boom in IT industries. As a business city, it repeatedly received the award 'Best City to Work in' by the Bombay-based magazine *Business Today*.[8] Against the background of the theoretical preliminaries, the features of a regional network in the 'high-tech' community of Bangalore will be presented, with some reference to the national development framework and evolutionary characteristics.

The Indian National Innovation System

There has long been a strong emphasis on technological development in India, first expressed in the Technology Policy Resolution of 1948, one year after achieving independence. But the trajectory was marked by a radical shift in direction from the paradigms of technological self-reliance and import substitution to those of liberalization and export promotion in the early 1990s. Despite the shift, the general approach that prevails is that of a planned economy (Misra and Puri, 1996). A lively discussion about the 'right way' is still going on today, focused on ways to better combine technological and industrial development (Pakrashi and Phondke, 1995; Krishna Murthy *et al.*, 1996). This indicates some shortcomings especially in the fields of transfer relations and the coordination of different technology actors. The principal concerns within the Indian innovation system will be illustrated below, referring to the nature of state activity, the orientation and objectives of R&D, and the innovative behaviour of industry.

The central state continues to strongly dominate the R&D activities in India, with the national Ministry of Science and Technology as well as the ministries of Defense, Electronics, and Telecommunication as major protagonists. The state is responsible for almost three-quarters of the entire national R&D expenditure of roughly US$ 2 billion (in 1994/95) – a very low sum in international comparison. Special emphasis has been put on R&D for defence sectors (e.g. electronics, aviation and space, and missiles), for general infrastructure support (e.g. atomic energy, electrical power, and communication), but also for basic science (astrophysics). Overall, then, the structure of real technological demand in the country is widely neglected (Chandrashekar, 1996). Accordingly there has always been a more mission-oriented than diffusion-oriented approach to technology policy, resulting in an extremely polarized picture: On one hand, India has won Nobel Prizes in science, develops nuclear devices, and launches its own satellites; on the other hand, industrial technology has to be judged more medieval than modern in wide areas of the country, fighting against the rigours of severe infrastructure constraints caused by an unchecked population growth.

The Indian government has installed an intricate network of support organisations, entailing a complex bureaucracy of interdependencies in decision-making and carrying out of public R&D. A central agency, the Council of Scientific and Industrial Research (CSIR), was established as early as 1942 with the task to promote and coordinate scientific R&D and industrial development. It has now grown to an 'organization of 40 laboratories and 80 field/extension/regional centres, including polytechnology transfer centres ... in the capitals of 7 States' (Joshi, 1995, p. 21). Though the CSIR also tries to promote national technology transfer, the specific regional opportunities of this strategy have not been discovered yet. A positive aspect about institution-building in India is that the state also follows a policy of considerably expanding the network of educational institutions, building upon some colonial heritage with the use of English as the official higher education language. The establishing of around a dozen – now highly reputed – Indian Institutes of Technology and Indian Institutes of Management as well as numerous polytechnical colleges provides a growing qualified workforce for sophisticated industries. The problem of an enormous brain drain to English-speaking industrial countries also must be mentioned.

As another sphere of state activity, PSEs have been given a crucial role in technology-based production. They were concentrated in particular in strategic, capital-intensive sectors, almost all of which can be subsumed under the 'high tech' label and show a high congruence with the foci of public R&D laboratories (Deolalikar and Sundaram, 1989). In India, PSEs and public R&D institutions have deliberately been linked together from the start, forming systems of collaboration and exchange in areas such as aircraft building, space technology, electronics, and telecommunication, in some cases also as spatial clusters. In spite of these complexes, the dependency on international technology purchases is high, for public as well as private firms.

The Indian private sector shows only a limited institutional ability to link up to centres of excellence. Companies have been quite passive in R&D so far, with only a quarter of the total Indian R&D input spent by the private sector (compared to almost 70% in South Korea; see Chandrashekar, 1996, p. 111) and an average R&D intensity of only around 0.8% (R&D expenses as a share of turnover) – and these figures represent only the fraction of firms even involved in R&D activities (Iyer, 1990). Though the number of firms carrying out R&D has increased in recent years – also due to the removal of restrictions to enter sectors previously reserved for the state – the majority still buys and uses technologically advanced equipment from outside rather than developing innovations themselves or in collaboration with national R&D institutions. After a long protection under a regime of import substitution, now the scenery changes in the wake of India's economic policy turn to outward liberalization in 1991. Incoming international competition gradually increases the pressure for local private as well as public firms to be technologically up-to-date. The state-promoted inflow of MNCs in knowledge-intensive sectors forms a particular new challenge to international as well as national technology collaboration in India.[9]

Empirical Base for Investigating the Technology Region of Bangalore and Its Networks

Since the question is to find out about people and their relational behaviour, a series of personal interviews in selected organizations was at the core of information collection, carried out in the first half of 1997. Both the cultural features of India and the focus of the project on relational behaviour made it necessary to visit all organizations personally and devote a great deal of time to the individual talks, limiting the number of objects to be included.

On the industry side, the top executives of 50 companies were selected as interview partners, with the firms representing the current large technology players in the area by the judgement of local experts. They provide a mixed sample of public (4%) and private (96%), foreign (50%) and Indian/joint venture firms (50%), permitting variations in contacting behaviour to be traced. Most firms belong to IT sectors with a bias towards software development (75%), but some other branches also were chosen to complete the picture (e.g. aircraft building and biotechnology). Finally, over 10 interviews were carried out with selected R&D institutes and organizations of industrial promotion in the region.

Because of the limited sample, there is no statistical representativeness of the study (the lack of detailed statistics on the industrial composition in the Bangalore urban district does not allow reliable calculations on it in any case). But regarding the central goal to identify networks of information exchange within the regional high technology community, the most important segment of actors were included. By using a specifically developed qualitative interview method called a 'holographic' interview,[10] it was also possible to capture more than just a single case package of information each time and get a wider and reconfirmed view on general developments in Bangalore.

Regional Characteristics of Bangalore

Since the successful development of an innovative region depends to a great extent on its historic heritage in knowledge accumulation (Conti, 1993), we also have to take a look at the history of Bangalore to understand its current attractiveness for technology-oriented activities. Decisive for its take-off, especially after 1991, have been the many educational institutions that Bangalore hosts, in combination with other factors forming a precondition for innovative regional development.

To begin with, Bangalore offers a high quality of life due to favourable climatic conditions. Its location almost 1000m above sea level justifies the label 'Air Conditioned City'. In addition, there are huge parks of British origin, leading also to the reputation as India's 'Garden City'.

Some important elements of the technological infrastructure were already established in colonial times, such as the Indian Institute of Science (1909), one of the most reputed institutions of R&D and higher education in the country, and a British base for aircraft maintenance (1940). After independence, the Indian government systematically built up the area as a centre of defense-oriented R&D and production, installing a network of 28 public research institutions in various fields (aeronautics and space, telecommunications, computing, artificial intelligence, and robotics) as well as technology-intensive PSEs.[11] The educational infrastructure also was substantially improved to provide the necessary workforce, with today 4 universities and over 300 colleges operating in the area.

Starting in the late 1970s and further eased by the Indian liberalization policy after 1991, an increasing inflow of foreign direct investment in computer hardware and software characterises the regional technological development. Attracted by the local skilled labour force and quality of life, most of the big global IT players now have Bangalore branches (such as Texas Instruments, Motorola, Hewlett Packard, Digital Equipment, IBM, Siemens, Novell, Oracle and many more). In addition, important Indian companies have grown up as well (e.g. Infosys, Wipro Systems, BFL Software, VXL Instruments). All together, around 70 – more than a quarter – of the 274 biggest software firms in India now are located in Bangalore, surpassing even the industrial capital, Mumbai (Bombay) with only 54 (Heeks, 1996).

Regarding the IT sector as a whole, rumours report 500 or more companies existing in the city, among them many small one-man shops. Over 160 firms (as of June 1997) are of significant size and substance, officially registered under the STP/HTP schemes (see note 9), making Bangalore a major concentration of IT industries even on a global scale.

There are quite a few efforts by central and state governments to support high-tech development. These include designating specific locational areas for IT firms such as 'Electronics City' and 'International Technology Park' in Bangalore suburbs. But the success of Bangalore is widely independent of these enclaves, with the STP status also available elsewhere (a mere 20% of the interviewed companies have or plan a (branch) location there). Instead, firms mostly give complaints and critique regarding the industrial areas and state policy in general, faulting the public inactivity to improve the 'cracking' infrastructure.

Networks of Relations in the Technology Region of Bangalore

In a regional development pattern very much influenced by exogenous forces recently, the question of the existence and nature of regional linkages of support and information exchange among technology protagonists of Bangalore is of immense interest. Two general observations should be pointed out:

- Substantial social networking takes place between technology-related actors in Bangalore; moreover, even most MNC branches – with their connections to global systems – are embedded in local relationships of various kinds, with personal factors playing an important role.[12]
- The nature and pattern of these linkages appear to be significantly different from those in technology regions of industrial countries, reflecting specific technological, industrial and cultural conditions in India as an only-recently-developing country.

A preliminary scheme of the regional network, consisting of several support systems, within actor groups as well as between them, can be outlined for the Bangalore example, illustrated in Figure 9.2 and Table 9.1. Within the relational scheme, three main actor groups, forming relatively coherent communities, can be distinguished:

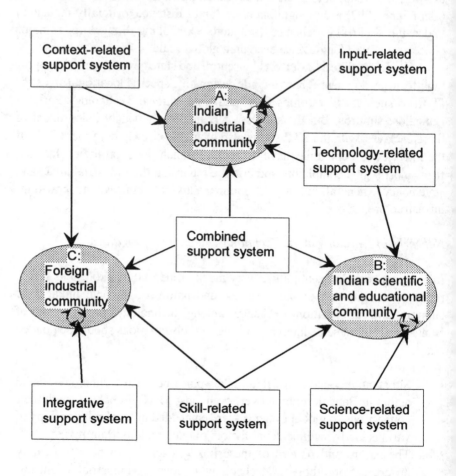

Figure 9.2 Support systems in the technology region of Bangalore

Table 9.1 Components of the support systems in Bangalore

Context-related support system	Informal industrial associations Formal industrial associations Exchange of workforce Border-crossing joint ventures Supply/subcontracting from local firms to MNCs Supply from MNCs to local firms Private community
Input-related support system	Supply/subcontracting linkages Exchange of workforce Technical consulting by PSE experts Collaboration of spin-off companies
Technology-related support system	R&D collaboration in different forms Exchange of workforce Supply of equipment
Integrative support system	Alliance of mother companies Locational consultancy Exchange of workforce Supply of equipment Private 'expat' community
Science-related support system	Close R&D collaboration Exchange of scientific staff Personal relationships
Skill-related support system	Recruitment of qualified people Upgrading of local higher education Donation of equipment Consulting/Further education to firms R&D collaboration and technology transfer Indian university graduates as CEOs
Combined support system	Joint ventures of PSEs with foreign MNCs

- The Indian industrial community (A) (consisting of private firms, joint ventures of Indian with foreign firms, and PSEs).
- The Indian scientific and educational community (B) (comprising the local institutions of R&D and higher education).
- The foreign industrial community (C) (formed by the branches of foreign MNCs).

Missing elements might be noticed, such as public administrative organisations and industrial promotion agencies. But in the case of Bangalore, they are not really seen as autonomous agents or major protagonists in the regional network, only as occasional linkage mediators.

The term 'community' indicates that in the overall system the support systems related to the internal group relationships dominate, being characterised by common interests and a sense of belonging of a majority of members.

Let us first discuss the important relations within each group of actors:

Input-Related Support System   The regional linkages within the Indian industrial community (A) can be characterized as an 'input-related support system', since they mainly are the outcome of a vertical quasi-integration of different firms in the value chain, still influenced by the former Indian policy of self-reliance. This linkage system shows some similarity to buyer-supplier networks in industrial economies, but the bad transport and telecommunication conditions in India have especially enforced their local clustering in an industrial district manner. In particular the PSEs stand out as 'producers' of a regional network of vendors consisting partly of spin-offs, due to their long locational tradition and big size. There is also the interesting observation of technical consulting by senior experts from PSEs to other Indian firms. This reflects the relatively high sophistication achieved in the public sector compared to the private one, on the one hand. On the other, it indicates the necessity of PSE employees to find additional sources of income due to the low wages in public firms and their recent ambitions to get rid of surplus employees in order to transform themselves from 'merit companies' to global competitors.

Science-Related Support System   The support system of the local scientific and educational community (B) highly resembles the research networks in industrial countries (not surprisingly given the highly international character of science in general). There are manifold collaborations taking place, often including the exchange of scientific staff on all levels. Accordingly, strong personal ties exist between regional R&D laboratories and to universities, resulting in an intense informal information exchange.

Integrative Support System Within the Bangalore community of foreign company branches (C) a linkage system is also growing, contradicting some assumptions about 'typical' MNC behaviour. It is labeled the 'integrative support system' since it integrates MNCs into the local environment. The main purpose is to reduce the locational uncertainty associated with the foreign location of Bangalore from each MNC's view. Being virtually neighbors not only allows the firms to easily transfer alliances of their mother companies (with headquarters mostly far away) to the regional scale, but it also makes 'competitors at home become friends here', as one company executive put it. This results in market transactions such as the supply of computer equipment to each other, though these are insignificant from the companies' perspective. Also, some 'soft' support takes place: MNCs give informal locational consultancy advice to incoming ones, and a private community of expatriate executives has emerged, facilitating the sometimes difficult business and private lives in the Indian environment. Another important factor is the extensive exchange of Indian workforce between MNC branches,[13] providing an intricate web of personal relationships on the employee level.

The inter-group support systems of Bangalore's technology actors are of specific interest and of growing significance on the regional scale, linking the scientific with the industrial sphere, the foreign with the local one, and enhancing regional knowledge transfer as well as self-reinforcing learning processes.

Technology-Related Support System Between the Indian industrial and the scientific and educational communities (A and B) there is a long tradition of interaction. The so-called 'technology-related support system' helps to develop indigenous technical know-how in R&D institutes for production and problem-solving in industry. But on the side of the Indian industrial community (A), the integration is generally limited to PSEs and to firms that have spun off from local laboratories. Regarding the type of linkages, this system again shows some resemblance to patterns in the industrial world, but with a tendency to circulate and process externally-purchased technology and know-how within the regional network. The forms of collaboration comprise the whole range of contracted or cooperative R&D and technology transfer, as well as the supply of equipment from companies to research institutes. Particularly noteworthy is a significant flow of graduates from the regional institutions of higher education to top positions in local firms, where these people usually are the agents for a continuing contact to their former place of study and to fellow students in other (local) organizations.

Context-Related Support System   Within the industrial sphere, between the Indian and foreign industrial communities (A and C), the proliferation of a foreign-local network can be observed. Since a major driving force is the IT companies' effort to improve the local – and often extremely bothersome – context of operation affecting all of them in Bangalore, it is labeled the 'context-related support system'. The most interesting expression of this connection is the formation of specific problem-solving associations, as a formalisation of actually informal negotiations: Organised by sectors (e.g. Bangalore Forum of IT) and by locations (Electronics City and Peenya Industrial Area associations), top executives from powerful foreign and Indian firms regularly meet to combine their strengths in fighting locational bottlenecks, such as severe infrastructure constraints, bureaucratic impediments or the enormous brain drain. Joint initiatives for socially beneficial work (for example, building schools) also are discussed frequently in these circles, and private personal communities emerge from them. In a similar way, meetings and common target negotiations in the course of active memberships in formal industrial associations of the region have a unifying effect (such as the Greater Mysore Chamber of Industry, Confederation of Indian Industry, and bilateral chambers). Besides these contact forums of the big regional players, smaller local firms also increasingly get in touch with MNC branches, mostly through the gradual build-up of supply or subcontracting relationships. In terms of knowledge flow and learning processes, the active raising of qualified local suppliers by MNCs is an especially interesting development, generating competence and long-term relationships of trust. This local networking tendency is connected with a techno-functional upgrading process of the Bangalore operations of many MNCs, accompanied by the disintegration of routine work. As between MNCs, there also is a considerable exchange of workforce between indigenous and foreign firms, in this case accelerated by wage differentials and prestige reasons.

Skill-Related Support System   Evidence could be found as well regarding an increasing interaction of foreign firms with Bangalorean scientific and educational institutions (B and C). The term 'skill-related support system' therefore relates to skilled people as the most important local resource of MNC branches. Accordingly, the recruitment of graduates from regional universities and colleges is at the core of this subsystem. But there is more than just a one-way flow, since a growing number of MNCs engage in upgrading local education in terms of state-of-the-art and application-oriented contents. This sometimes even includes lecturing by top executives at universities or the donation of sophisticated equipment. Though the companies have quite egotistic motives for doing this, the results of this 'beneficial egotism' are very positive for the region as a whole. Other kinds of collaboration emerge from this, such as R&D partnerships and technology transfer, with information flows mostly taking the opposite direction common in the industrial countries: from company to laboratory/ university, where they get further disseminated. For management/technical consulting and further education, some MNCs also are using expertise from regional institutes. Within this very interesting and dynamic contact system, a specific group of people plays a decisive role: The majority of regional MNC subsidiaries is not headed by foreign, but by Indian executives, with some of them even coming from the Bangalore region. Though most of them – after acquiring a first degree in India – studied and worked abroad (sometimes with the same company sending them back to put up the Bangalore branch), they still have strong social and emotional ties to their home country, are delighted to enter local personal communities, and already know the local R&D scene quite well. Some of them still have good contacts to former places of study in the area, making collaboration easy.

Combined Support System   We finally come to the support system combining and linking all three Bangalore technology communities (A, B and C). It rests on a single observation, but one that suggests an emerging trend with wide possible impacts. There is a growing number of joint venture companies established by technology-oriented PSEs with foreign partners, integrating the latter into the traditionally close regional collaboration system of the state firms with Bangalore's scientific and educational community.

Conclusion

Presenting the case of the Indian technology region of Bangalore and a model of its internal network of technological actors, this chapter has attempted to contribute to a set of three main research issues:

The first issue is the extent to which developing countries like the NICs have a chance to experience a sustained growth of technology regions based on local relationships of innovative stimulation and support, which characterizes successful regions in developed countries. Looking at the major technological development characteristics of NICs shaping their national innovation systems, these countries face specific impediments in the formation of national/regional knowledge-intensive linkages, mainly because of a dominance of foreign know-how and investment and a lack of diffusion-oriented policies. But many countries already have taken the opportunity to gradually improve their conditions by building local technological capabilities in R&D-institutions and industry. Regional systems of technology collaboration can also emerge in advancing NICs, but they depict patterns significantly distinct from the ones to be found in industrial countries.

The second issue is represented by the Bangalore case as a new type of (unplanned) technology region different from examples in the industrialized world. This empirical example demonstrates that within the Indian national innovation system some regional networking takes place that is reminiscent of the creative milieus or industrial districts identified in industrialised countries. But the linkages found in Bangalore are specifically shaped by the general national development framework, as assumed in the theoretical part of this chapter.

The third issue deals with the nature of the regional relations actually constructing a milieu- or district-like fabric of mutual encouragement, support and social coherence towards innovation-based economic development. The Bangalore example shows that, in addition to formal linkages, more emphasis has to be put on informal, personal relationships occurring routinely alongside daily operations.

All this is more good than bad news for developing countries. It demonstrates that also for them the proliferation and 'trickling-down' of technological impulses can follow (pre-existing) regional contact systems. But the task remains to find the right political promotion measures, since these systems emerge by themselves and can hardly be constructed.

Acknowledgements

The chapter derives results from ongoing research, funded by the Deutsche Forschungsgemeinschaft (the German National Science Foundation). The committed assistance of my husband Günter Eisebith to the empirical work was a major key to its success, enabling us to get integrated into regional personal networks ourselves. Thus the following results owe a lot to his participation, for which I thank him very much. A special thanks also goes to our local partner and important contributor to the project, the Greater Mysore Chamber of Industry and secretary Mr. Vishwanathan. Finally, the author thanks the editors of this book for their useful comments.

Notes

1   The definition of 'technology region' used here after Sternberg (1995, p. 7) is the following: A region below national level, being characterized by a proportion of firms and employment in high-tech sectors far above national average. With reference to less developed countries showing a very low overall average it should be added: above the high-tech average of industrial countries.

2   In this chapter the term 'network' is used in a general sense, as 'a configuration of nodes and links; that is, a system in which some elements are connected to other elements in a systematic way' (Batten *et al.*, 1995, p. vii).

3   Some authors leave aside this aspect of social coherence when assigning the label 'industrial district' to any spatial cluster of economically specialized firms, no matter if socially connected or not (as Park and Markusen, 1995). This broad conception of an industrial district is explicitly excluded in my chapter.

4   Whether a country is regarded as a NIC today seems to be left to individual judgement, since no common characteristics are used by different authors or organizations. But it is widely agreed on that the group of the so called first-tier NICs (South Korea, Taiwan, Hong Kong and Singapore) is now followed by another group of second-tier NICs, comprising countries such as Malaysia, Thailand, Indonesia and India as well as Brazil, Mexico and Argentina (Chowdhury and Islam, 1993).

5   Prominent examples, besides the Indian one examined below, are for instance the regions of Taejon (South Korea) and Penang (Malaysia).

6   This should not be misinterpreted: There is no doubt about the general necessity and benefit of appropriate technology for developing countries. On a lower level of sophistication, they even form a prerequisite for allowing local linking of firms. But regarding high-technology, the developing countries' emphasis on appropriate technologies is a disadvantage.

7   There are other reasons for NICs to be favoured as locations of new innovative firms: High-tech companies show a tendency to locate in areas not marked by previous industrial concentrations (Storper and Walker, 1989), though certain basic requirements regarding local labour and know-how have to be fulfilled. This gives a chance to '(semi)peripheral' regions and countries.

8   Issues from December 22, 1994, and December 22, 1996.

9   A whole set of policy instruments is particularly luring computer and software firms to India, such as an infrastructure of industrial areas providing satellite communication links and incentive programmes releasing the firms from restrictions and obligations (e.g. Software Technology Park (STP) and Hardware Technology Park (HTP) status).

10   The term refers to two significant characteristics of a holograph: It is a multi-dimensional picture having the quality that, broken into pieces, each fragment still contains the whole image.

11   The most important ones are Indian Telephone Industries (1948), Hindustan Aeronautics Ltd. (1951), Bharat Electronics Ltd. (1954) and Hindustan Machine Tools (1955).

12   The piecemeal build-up of local linkages can also be demonstrated by examples from other Asian NICs (e.g. Park, 1996; Rasiah, 1994), nurturing the optimistic assumption that technological trickling-down effects are increasingly possible for less developed countries.

13   Recently an average annual attrition rate of 25% was experienced in the software sector of Bangalore, though only a small fraction of people change to other regional employers. The majority leave to search their luck in the U.S.

References

Amin, A. and Thrift, N. (1994), 'Living in the Global', in A. Amin and N. Thrift (eds), *Globalization, Institutions, and Regional Development in Europe*, Oxford University Press, Oxford, pp. 1-22.

Baranson, J. (1969), 'Role of Science and Technology in Advancing Development of Newly Industrializing States', *Socio-Economic Planning Sciences*, vol. 3, pp. 351-383.

Batten, D., Casti, J. and Thord, R. (eds.) (1995), *Networks in Action: Communication, Economics and Human Knowledge*, Springer, Berlin.

Bell, M. and Pavitt, K. (1993), 'Technological Accumulation and Industrial Growth: Contrasts between Developed and Developing Countries', *Industrial and Corporate Change*, vol. 2, pp. 157-210.

Brugger, E.A. (1990), 'Endogenous Development between Myth and Reality: Pre-Requisites for Endogenous Development Strategies', in: H.-J.Ewers and J. Allesch (eds), *Innovation and Regional Development*, Springer-Verlag, Berlin, pp. 161-166.

Brunner, H.P. (1995), *Closing the Technology Gap: Technological Change in India's Computer Industry*, Sage, New Delhi and Thousand Oaks, CA.

Buckley, P.J. and Clegg, J. (eds) (1991), *Multinational Enterprises in Less Developed Countries*, Basingstoke, Macmillan.

Camagni, R. (ed) (1991), *Innovation Networks: Spatial Perspectives*, Pinter, London.

Camagni, R. (1995), 'Global Network and Local Milieu: Towards a Theory of Economic Space', in S. Conti, E.J. Malecki and P. Oinas (eds), *The Industrial Enterprise and Its Environment: Spatial Perspectives*, Avebury, Aldershot, pp. 195-214.

Cantwell, J.A. (ed) (1993), *Transnational Corporations and Innovatory Activities*, United Nations Library on Transnational Corporations, London.

Casson, M. (ed) (1991), *Global Research Strategy and International Competitiveness*, Basil Blackwell, Oxford.

Chandrashekar, S. (1996), 'Technology Policy and Economic Reform – The New Imperatives', in M.V. Krishna Murthy, N.S. Siddharthan and B.S. Sonde (eds), *Future Directions for Indian Economy: Technology Trade and Industry*, New Age Intl. Ltd. New Delhi and Bangalore, pp. 97-116.

Chowdhury, A. and Islam, I. (1993), *The Newly Industrialising Economies of East Asia*, Routledge, London.

Conti, S. (1993), 'The Network Perspective in Industrial Geography. Towards a Model', *Geografiska Annaler*, vol. 75B, pp. 115-130.

Conti, S., Malecki, E.J. and Oinas, P. (eds) (1995), *The Industrial Enterprise and Its Environment: Spatial Perspectives*, Avebury, Aldershot.

Conti, S. (1997), 'Technological Change in Space. An Introduction', Paper presented at the Residential Conference of the IGU Commission on the Organisation of Industrial Space, Göteborg, Sweden, 4-8 August.

Delaplace, M. (1993), 'High tech et facteurs de localisation: une revue de la littérature', *Revue d'Économie Régionale et Urbaine*, no. 4, pp. 679-704.

Deolalikar, A.B. and Sundaram, A.K. (1989), 'Technology Choice, Adaptation and Diffusion in Private- and State-owned Enterprises in India', in: J. James (ed.), *The Technological Behaviour of Public Enterprises in Developing Countries*, Routledge, London, pp. 73-138.

Dicken, P. (1992), *Global Shift: The Internationalization of Economic Activity*, 2nd edition, Paul Chapman, London.

Dunning, J.H. (1993), *Multinational Enterprises and the Global Economy*, Addison-Wesley, Wokingham.

Enos, J.L. (1991), *The Creation of Technological Capability in Developing Countries*, Pinter, London.

Esser, K., Hillebrandt, W., Messner, D. and Meyer-Stamer, J. (1992), *Neue Determinanten internationaler Wettbewerbsfähigkeit - Erfahrungen aus Lateinamerika und Ostasien*, Deutsches Institut für Entwicklungspolitik, Berlin.

Foray, D. and Freeman, C. (eds) (1993), *Technology and the Wealth of Nations. The Dynamics of Constructed Advantage*, Pinter, London.

Fröbel, F., Heinrichs, J., and Kreye, O. (1986), *Umbruch in der Weltwirtschaft. Die globale Strategie: Verbilligung der Arbeitskraft, Flexibilisierung der Arbeit, Neue Technologien*, Rowohlt, Reinbek.

Fromhold-Eisebith, M. (1995), 'Das "kreative Milieu" als Motor regionalwirtschaftlicher Entwicklung', *Geographische Zeitschrift*, vol. 83, no. 1, pp. 30-47.

Heeks, R. (1996), *India's Software Industry: State Policy, Liberalisation, and Industrial Development*, Sage, Thousand Oaks, CA.

Holmström, M. (1994), *Bangalore as an Industrial District: Flexible Specialization in a Labour-Surplus Economy?* Pondy Papers in Social Sciences no. 14, French Institute Pondicherry, Pondicherry.

Iyer, P.P. (1990), *R&D in Indian Industry: A Selective Survey*, Technical Report, Centre for Scientific and Industrial Consultancy, Indian Institute of Science, Bangalore.

James, J. (ed.) (1989), *The Technological Behaviour of Public Enterprises in Developing Countries*, Routledge, London.

Joshi, S.K. (1995), 'Impact of Liberalisation on CSIR', in S.C. Pakrashi and G.P. Phondke (eds), *Science Technology and Industrial Development in India*, Wiley Eastern, New Delhi, pp. 21-29.

Krishna Murthy, M.V., Siddharthan, N.S. and Sonde, B.S. (eds) (1996), *Future Directions for Indian Economy: Technology Trade and Industry*, Proceedings of the Workshop at Bangalore, 19-20 Jan. 1996, New Age Intl. Ltd., Delhi and Bangalore.

Kuemmerle, W. (1997), 'Building Effective R&D Capabilities Abroad', *Harvard Business Review*, vol. 75, March-April, pp. 61-70.

Lall, S. (1992), 'Technological Capabilities and Industrialization', *World Development*, vol. 20, pp. 165-186.

Malecki, E.J. (1997), *Technology and Economic Development. The Dynamics of Local, Regional and National Competitiveness*, 2nd edition, Addison Wesley Longman, London.

Marton, K. (1986), *Multinationals, Technology, and Industrialization: Implications and Impact in Third World Countries*, Lexington Books, Lexington, MA.

Misra, S.K. and Puri, V.K. (1996), *Indian Economy*, 14th edition, Himalaya Publishing House, Bombay.

Nadvi, K.M. (1992), *Flexible Specialization, Industrial Districts and Employment in Pakistan*, ILO Working Paper no. 232, International Labour Office, Geneva.

Nadvi, K.M. (1994), 'Industrial District Experiences in Developing Countries', in *Technological Dynamism in Industrial Districts*, United Nations, New York, pp. 191-255.

Nelson, R.R. and Rosenberg, N. (1993), 'Technical Innovation and National Systems', in R.R. Nelson (ed.), *National Innovation Systems: A Comparative Analysis*, Oxford University Press, Oxford, pp. 3-21.

Oinas, P. (1997), 'On the socio-spatial embeddedness of firms', *Erdkunde*, vol. 51, no. 1, pp. 23-32.

Pakrashi, S.C. and Phondke, G.P. (eds) (1995), *Science Technology and Industrial Development in India*, Wiley Eastern Ltd., New Delhi.

Park, S.O. (1996), 'Networks and Embeddedness in the Dynamic Types of New Industrial Districts', *Progress in Human Geography*, vol. 20, pp. 476-493.

Park, S.O. and Markusen, A. (1995), 'Generalizing New Industrial Districts: A Theoretical Agenda and an Application from a Non-Western Economy', *Environment and Planning A*, vol. 27, pp. 81-104.

Perez, C. (1985), 'Micro-electronics, Long Waves and World Structural Change: New Perspectives for Developing Countries', *World Development*, vol. 13, pp. 441-463.

Perez, C. and Soete, L. (1988), 'Catching Up in Technology: Entry Barriers and Windows of Opportunity', in: G. Dosi, C. Freeman, R. Nelson, G. Silverberg and L. Soete (eds), *Technical Change and Economic Theory*, Pinter, London, pp. 458-479.

Pyke, F. and Sengenberger, W. (eds) (1992), *Industrial Districts and Local Economic Regeneration*, International Institute for Labour Studies, Geneva.

Rasiah, R. (1994), 'Flexible Production Systems and Local Machine-Tool Subcontracting: Electronics Components Transnationals in Malaysia', *Cambridge Journal of Economics*, vol. 18, pp. 279-298.

Sachwald, F. (1995), 'The Organization of Technological Activities within Firms and between Countries: the Consequences of Globalization', in Tokyo Club Foundation for Global Studies et al.(ed.): *Tokyo Club Papers* No. 8, Tokyo, pp. 29-67.

Schamp, E.W. (1993), 'Industrialisierung der Entwicklungsländer in globaler Perspektive', *Geographische Rundschau*, vol. 45, pp. 530-536.

Schmiegelow, M. (1990), 'The Asian Newly Industrializing Economies: A Universal Model of Action', *Civilisations*, vol. 40, no. 1, pp. 133-170.

Schmitz, H. (1990), *Flexible Specialisation in Third World Industry: Prospects and Research Requirements*, New Industrial Organisation Programme Discussion Paper no. 18, International Labour Office, Geneva.

Schmitz, H. and Musyck, B. (1994), 'Industrial Districts in Europe: Policy Lessons for Developing Countries?', *World Development*, vol. 22, pp. 889-910.

Silveira, M.P.W. (ed.) (1985), *Research and Development: Linkages to Production in Developing Countries*, Westview Press, Boulder.

Soesastro, H. and Pangestu, M. (eds) (1990), *Technological Challenge in the Asia-Pacific Economy*, Allen & Unwin, Sydney.

Storper, M. and Salais, R. (1997), *Worlds of Production*, Harvard University Press, Cambridge, MA.

Storper, M. and Walker, R. (1989), *The Capitalist Imperative: Territory, Technology, and Industrial Growth*, Basil Blackwell, New York.

Sternberg, R. (1995), *Technologiepolitik und High-Tech-Regionen -- ein internationaler Vergleich*, LIT, Münster.

Stöhr, W. (1986), 'Regional Innovation Complexes', *Papers of the Regional Science Association*, vol. 59, pp. 29-44.

10 On Technology and Development

Edward J. Malecki, Päivi Oinas and Sam Ock Park

Making Connections: A Summary

This book has been about technological change and its role in (regional) development. One of the themes that comes up in various forms in the chapters is the importance of making connections: connections that enhance the performance of firms and their local surroundings. The crucial distinction that this book emphasises is the one between *local* and *nonlocal* connections. Recent thinking has incorporated the idea of an ongoing interaction of global and local networks, in which actors are a part (e.g., Amin and Thrift, 1992; Conti, 1993). However, past research on industrial districts, and the discussion it has launched in economic geography and related fields, has stressed primarily the essential nature of local relations and the kinds of conditions that seem favorable for the functioning of local production and innovation networks. This emphasis on local connections has caused many to overlook the critical role of links to nonlocal networks (as we point out in Chapter 1). Nonlocal networks provide access to knowledge not available in the locality, such as scientific discoveries and best-practice technologies, and they may facilitate incremental innovations in technologies already in use.

Two levels of connectedness or connectivity stand out in the preceding chapters of this volume: the one that has to be formed and maintained by *firms* and the one that is formed at the level of entire *regional economies* in order to either 'get in the loop' – or to stay there. At times, these local connections are made collectively, with the intention of creating a successful regional economy with favorable conditions for learning to take place in and among firms and the various support organisations. At other times, the connections are made more independently by several actors but, even if not intentionally coordinated, jointly they may comprise an essential array of associations that strengthen the entire social and economic fabric of the region.

Local Relations

The contacts that firms make and maintain with others provide them – and thereby their local and regional territories – with new knowledge that can be put to use in new products and processes, and in marketing existing products and services in new markets. This is done in many ways, some of them involving codified, public knowledge such as publications. A great deal of the most useful information, however, is available only from face-to-face contact, for it is through face-to-face contact that tacit knowledge and knowledge based on learning-by-using can be communicated, even in this age of telecommunications (De Meyer, 1993; von Hippel, 1994). Such close contact is still constrained by the friction of distance: it is easier to make contacts locally than at a distance – especially casual, unplanned and social contacts, from which useful knowledge also may result.

Knowledge flows, learning and innovation are facilitated by the greater interaction between actors. Close regional interactive relations between various local actors may evolve into local or regional 'enterprise support systems' which are strengthened by associative governance systems where private bodies take some responsibility for policy development and administration (e.g. Chambers of Commerce, regional conferences, consortia and other arrangements for interaction at local and regional levels). These are found in the many European examples which Asheim and Cooke provide, but similar support systems can be identified, e.g., in India (Fromhold-Eisebith, Chapter 9).

Nonlocal Relations

In the absence of a supportive local milieu, firms rely on sources of technology from national and international networks of customers and suppliers and other third parties. Alderman's findings (Chapter 4) on this reinforce earlier research (Malecki and Veldhoen, 1993; Vaessen and Wever, 1993; Vatne, 1995). Indeed, it appears that strong local networking may be less common – and less necessary – than has been thought. Instead, close links to suppliers, customers and others – wherever they are located – are vital to a firm's competitiveness.

Nonlocal networks, however, bring along challenges. Not only must firms know what their competitors are doing, they also must have a presence where they have important customers so that they can know what those customers want, and are able to serve their needs. Despite global competition, culture continues to make its presence felt: some products, e.g., complex machinery, are not used or understood in the same way in different places (Gertler, 1995, 1996). The same is true of specialised or advanced producer services, on whose expertise manufacturers must draw upon. Advanced producer service firms also provide links to global networks, centrally since they 'participate in networks embedded in systems of major cities where their most important markets are to be found' (Moulaert *et al.* 1997, p. 105). Indeed, one of several roles of cities is to provide formal and informal socialisation at the local level in what Crevoisier calls 'interaction and learning sites' (Chapter 3; see also Amin and Thrift, 1992; Lambooy and Moulaert, 1996). Because knowledge that can only be accessed through external relations might require a 'technological code', there is a need for translators or gatekeepers who have an ability for 'versatile integration', coding and decoding knowledge from many different sources (Becattini and Rullani, 1996).

Nonlocal links operate in two directions. Just as companies keep up with events and activities of their counterparts elsewhere, so too do vigilant competitors attempting to gather product, market, and production knowledge in other places. This is part of the ubiquitification process described by Maskell (Chapter 2), in which formerly unique and localised inputs – both material and nonmaterial (knowledge, ideas, concepts, designs) – may become 'ubiquities'. To counteract this process, local actors evolve inter-organisational routines for knowledge creation and transfer. These routines increase the efficiency of the economy, not only by lowering transaction costs – the costs of persuading, negotiating, co-ordinating, understanding and controlling each step in a transaction between firms – but also by enhancing transaction benefits with their provision of 'thick' information (Imai and Baba, 1991; Maskell *et al.*, 1998; Sako, 1992).

Close producer-user interaction has enabled some firms to exploit niches and to carve out strong positions in some markets. Large firms (LFs), which are small in global terms, are sometimes best placed to derive benefits from their 'local' resources and customers. Patchell, Hayter and Rees (Chapter 5) show that LFs are able to incrementally create strong, even dominant, positions in their market niche. By building and maintaining close ties to innovative customers in international markets, LFs are able to become successful producers and suppliers. Also, the close ties with foreign customers maintained by the British engineering firms, studied by Alderman (Chapter 4), keep them competitive despite the absence of strong locally based networks.

Translating the Network Model for Less Developed Countries and Regions

At present, much contemporary development thinking, based largely on experience in Western economies, involves an emphasis on the significance of tightly collaborating network relationships in a strong and supportive local milieu (see, e.g., Maillat, 1998). At its best, such a milieu embodies cultural creativity enabling learning and leading to progress being made by incremental or radical innovation. The greatest potential for this to occur is in regions where 'there is systemic, i.e., regular, two-way, interchange on matters of importance to innovation and the competitiveness of firms' (Cooke *et al.*, 1998, p. 484). Even in regions where local institutional support is weak, industrial specialisation agglomerates around the innovative strength of the local area. It remains difficult to understand exactly what makes a 'learning economy' arise in some places and not in others (cf. Eskelinen and Kautonen, 1997; and Maskell, 1997; and the variation found in the various 'regional innovation systems' in Braczyk *et al.*, 1998).

Developing countries, especially in Asia, recognise that new-tech and high-tech industries can be important elements of an economy and a means of developing human resources in other than low-wage, routine manufacturing. Connecting to the new industries requires determined effort, as several chapters in this volume have illustrated. The connections can be a result of inward movement, as in the case of investment and activities of global firms or MNCs, or it can take place through outward search and learning, including education abroad, travel and other global search mechanisms (Park and Markusen, 1995). In China and Korea, e.g., many such networks are at first predominantly nonlocal, based on connections to MNCs, and only after spinoffs and small local firms emerge have local networks intensified. In general, it is entrepreneurship that most enhances local linkages and other aspects of integration in new industrial districts in East Asia (Park, 1997).

MNCs provide vital connections to the global economy, despite the fact that foreign investment is not always looked upon as a positive element. These connections are critical for gaining knowledge about product markets and standards, as well as about new technologies. Outward search and learning includes travel abroad, such as the trip to Silicon Valley made by the Chinese officials that Wang describes (Chapter 8) or education and employment abroad by native Indians who returned to Bangalore, as Fromhold-Eisebith reports (Chapter 9). Travel abroad has long been a means of learning by late-industrialising countries (Myllyntaus, 1990). Links to foreign firms in

Zhong'guancun, Wang shows, provide capabilities for some enterprises that could only be obtained in this way. In both the Chinese and Indian cases, strong local research institutes and universities raise the 'absorptive capacity' of the region for external technology, that is, R&D and other activities associated with technological development such as technology adoption (Cohen and Levinthal, 1990).

A dilemma for developing countries is the desire not to be exploited by MNCs. Experience shows that global firms hold the power in bargaining with host countries and areas (Ruigrok and Van Tulder, 1995; Dicken, 1998). The influx of MNCs into the former socialist economies in Central and Eastern Europe has aroused fears of further marginalisation. Transition is still taking place and, as Barta (Chapter 7) points out, the lack of understanding of the 'rules' of capitalism – and how they could and should be applied to fit the transition economies and societies at large – has been a key problem. There is confusion concerning 'what kind of capitalism' should be aimed at. In Hungary and elsewhere in Eastern Europe, investment decisions by MNCs are seen as not entirely beneficial for local economies, yet the connections to global markets are essential for economic growth. The findings Barta reports suggest that the joint ventures and other foreign-owned operations in Hungary have been overly concerned with control. Zeira *et al.* (1997) suggest that the most successful joint ventures in Hungary are those in which neither partner is dependent on the other for most of their resources. Radošević (1997) sees opportunities for economies in transition to close gaps in capabilities from all types of international links, including subcontracting. Although it provides rather low benefits regarding management, subcontracting improves product development and market development.

The state tends to play a larger role in transitional and developing countries than in Western countries. Sometimes this is a result of bureaucratic networks that have only one-way, top-down links. In such cases, few local networks can form, resulting in what Grabher (1994) calls a 'disembedded regional economy'. The institutional changes necessary to support a productive economy which learns new capabilities requires that firms and institutions *unlearn* old ways (Choo, 1996; Johnson, 1992). In order to change the many 'obsolete competencies', change must go deeper than merely changes in institutions and incentive structures to the development of in-house capacities for learning and improvement (Pavitt, 1997). Part of what may need to be 'unlearned' is the lack of trust that often accompanies hierarchical relations, to be replaced by collaboration and shared knowledge (Lütz, 1997).

Constructing trust is a process that takes time, but its payoff can be large (Humphrey and Schmitz, 1998).

Forming local networks and the supporting *milieus* around them tends to be difficult. Wang (Chapter 8) shows the difficulty faced in China as that country attempts to create industrial districts. Local networking has begun, based on spinoff enterprises from university departments and research institutes, but they are not very strong. External contacts with foreign firms provide many of the new ideas, experiences, standards and product specifications that mark competitive products (Gu, 1996), and help policymakers in the structuring of institutions to support adopting, and increasingly adapting, new technologies and developing innovative activities of firms in their territory. In the following, we turn to such policy considerations.

Facilitating Regional Connections

There is a current tendency to believe that the private sector is better than the state-owned sector at making decisions regarding competitive enterprise. Yet there remains a role for the state in several matters, such as the provision and maintenance of infrastructure, education, research, training, and various support activities, either independently or in collaboration with private actors. These activities themselves need connection to local circumstances – or context-specific 'translation' – to be useful in different places.

If national policies are relatively less important than in the past, regional and local policies – and regional and local systems of innovation – have risen in significance. Some of this takes place through the actions of 'community entrepreneurs' or gatekeepers, including policymakers who encourage inter-firm communication and cooperation. The presence of local 'community entrepreneurs' – people with local knowledge and an ability to tap into local resources – contributes greatly to the kinds of circumstances that Amin and Thrift (1993) call institutional thickness. At the same time, there is a risk of 'institutional overkill' – the creation of an isolated organisation for every need. This is what, e.g., MacLeod (1997) has found in Scotland, in areas where the potential impact of support organisations has been greatly reduced by their failure to network with each other.

Policies aim at creating local and regional systems of innovation in localities and regions where they have not emerged (Camagni, 1995). The infrastructure and institutions address multi-faceted needs, including technical education and services for small firms. The experience of industrial districts suggests that policies aimed at maintaining a high level of information flow and worker skills, through training and technology upgrading, can be very effective. Regular interaction between government actors and those in the private sector in a locality – the local corporate-government interface – appears to be of considerable importance for the creation of a supportive milieu for new firms.

The local technological infrastructure is often both difficult for local policies to influence and difficult to measure (Autio, 1998). Local policy support or 'sponsorship' of entrepreneurship contributes significantly to the rich infrastructure in Linköping, Sweden (Asheim and Cooke, Chapter 6; Jones-Evans and Klofsten, 1997). However, the model represented by Silicon Valley, with its succession of spinoff firms, has inspired an assortment of policies to boost regional innovative capacity and entrepreneurial activity. Sternberg (1996) concludes that the implicit effects of technology policies that lack explicitly spatial goals dominate and result in spatial unevenness, reinforcing a few core regions. This generalisation holds across a range of countries with different policy traditions.

Based on an investigation of European industrial districts, Humphrey and Schmitz (1996) conclude that policies did not create, but followed and supported, the districts. They suggest that three C's were at work: the districts were customer-oriented, collective, and cumulative – much the same as the advantages of Silicon Valley. There appears to be a need for policies to support small firms, whose resources are too limited for effective scanning and learning. While networks are needed for the sharing of information, public provision of information seems to be essential for its initiation (Chabbal 1995; Estimé *et al.*, 1993). Yet even information provision is increasingly done on a private basis: specific firms specialise in, e.g., Internet-based information services in specific fields.

We know that small firms (SMEs) are important in spatial innovation systems, but we do not know how best to stimulate change in these firms. Left on their own, small firms prefer to stick to their routines and, even with new information in hand, may fail to act on it (Glasmeier *et al.*, 1998). The policy orientation (top-down or bottom-up) matters a great deal, because differences in national systems of innovation are reflected in large-scale differences in technology transfer programmes (Hassink, 1992). Policies that are aimed principally at small firms also may fail, unless they include both a proactive process of seeking out small-firm clients (rather than waiting for active firms to seek out information) and long-term commitment and funding to provide continuity (Hassink, 1996). Finally, the organisations that assist small businesses should be entrepreneurial in nature themselves in order to respond flexibly to the differentiated needs of local environments (Gibb, 1993; Johannisson, 1993).

Technology policy has traditionally been predominantly inward-looking. A set of policies to *create* technological capability, manifested in industrial policies and in science and technology (S&T) policies, attempts to support R&D and to 'target' certain industries or technologies that are considered to have especially high potential for future growth. These priorities are based on the linear model of innovation (as discussed in our Chapter 1 and by Asheim and Cooke in Chapter 6), and rely on a top-down approach rather than on the learning model that sees benefits in widespread sources of knowledge and incremental improvement. The inward focus is found with variations related to the deeply rooted characteristics based in culture that affect the policies of countries. One manifestation of culture is seen in the distinction between 'diffusion-oriented' and 'mission-oriented' policies, and in the effects of these policies on technology sharing and institutional structures. *Diffusion-oriented* policies 'seek to provide a broadly based capacity for adjusting to technological change throughout the industrial structure' (Ergas, 1987, pp. 205). *Mission-oriented* policies, on the other hand, have a top-down focus on a particular mission or objective, such as energy or national defense. The USA, the UK, and France have followed mission-oriented policies, and their political and industrial structures have inhibited technology and knowledge transfer. Germany, Japan, Switzerland, and Sweden, on the other hand, have been diffusion-oriented, with more sharing of technologies within as well as among firms (Chesnais, 1993; Walker, 1993; Ziegler, 1992).

The complex interlinkages among policies necessary for a diffusion-oriented system represent an institutional 'technology infrastructure' provided by a variety of institutions, public and private, with its principal objectives to create capabilities and build markets for new technologies (Justman and Teubal, 1995). This broader notion of a dynamic *innovation policy*, 'focused on the promotion and adoption of new technology (that is, the commercial development of the fruits of basic research)', recognises the corporate–government interface as an essential aspect of innovation policy (OECD, 1995; Ostry, 1990, p. 53). To these explicit policies, we could add the development of conventions that support the notion of change, technological and otherwise.

What is usually included in the local technological infrastructure are the 'hard' elements, such as education and transportation and communication infrastructure, university and industrial R&D, agglomerations of related industry, and specialised business services (see, Feldman and Florida, 1994). These are typically difficult for localities and regions to change directly, but they can be influenced indirectly by investment and promotion of quality-of-life areas or amenities such as arts and culture to attract workers in knowledge-based activities. More difficult to modify and to measure are the usefulness and information content of interactions, though these may be critical (Autio, 1998; March Chorda, 1995).

The 'hard' aspects of technological capability have proved difficult enough for policy to deal with, enough so that we can ask how policy makers can ever deal with the 'soft' aspects – especially across borders. The soft aspects concern the creation of conventions (e.g., Storper, 1997; Storper and Salais, 1997) favorable to innovative activities, and possibly also the adoption of 'cultural models' for carrying out industrial activities from elsewhere. Also this depends on the region's connections to sources of ideas and knowledge outside the region. Among the local conventions or practices that varies a great deal from place to place is the tendency to communicate with others. Such a tendency (or its absence) is also related to the policy orientation at the national and other scales.

Few attempts have been made to characterise developing countries in this way. Both China and India, however, fit the pattern of a mission-oriented policy environment. There, the inward focus has resulted in a failure to be competitive, especially in high-technology industries (Bowonder, 1998). However, in both India and China there is evidence that some regions have formed the institutional and social links necessary for a competitive response. Fromhold-Eisebith (Chapter 9) illuminates the exceptional web of interactions that form the various 'support systems' for technology-based development in the Bangalore region, which are grounded in long-held conventions of the region. As a result of the support systems, the region is able to rise above the constraints that are characteristic of most Indian regions.

The needed 'soft' capabilities – the creation and/or adoption of suitable conventions in specific regional, sectoral contexts of innovation – need to be internalised at the level of individuals and the organisations in which they operate. Also organisational cultures differ in the degree to which they are able to support processes of learning (Schein, 1996). This is why policy has difficulties supporting the 'soft' elements of innovative capabilities – but suggests that proactive efforts and public-private collaboration might work best in changing established or routinised practices at the level of firms and their interaction.

Final Thoughts

The authors of this book have addressed the issue of 'making connections' differently, but all of them would acknowledge that linkages – local and nonlocal – are essential for the competitiveness of firms and the development of places. Interestingly, it has become clear that it is simply human social interaction that is the basis of many, if not all, competitive local and regional environments. Interaction and personal relationships remain to a large degree constrained by distance, regardless of Internet, but there are means to overcome the constraints of distance. However, it is not yet clear to us which kinds of processes or activities of innovation are dependent on proximity, i.e., constrained by the need to establish close personal relations at close distance in specific institutional and conventional set-ups, and which are those that can be carried out over long distances.

Returning to the spatial innovation systems framework of Chapter 1, we propose that there is a need to consider the following challenges in future research on technological development:

1) *Gain deeper understanding on networks.* There is a need to move beyond the simple statement of the need for firms to be networked: we need to understand the various types of networks that firms create for different strategic purposes. While various types of 'implementation networks' – and the regional environments that support them – might be highly important for firms to succeed in their respective competitive contexts, various types of 'learning networks' are likely to be more relevant for the competitive success of firms – and their regional environments – in the long run (Oinas and Packalén, 1998).

2) *Incorporate both the 'hard' and 'soft' aspects into explanations of (successful) technological change*: The research community at large is now trying to solve problems related to the 'soft' aspects of the economy: both their conceptualisation *and* their interconnection to the 'hard'.

3) *Understand the collective nature of technological development*: Collective action often has no formal manifestation, rendering research difficult. The important conventions in each place are the often informal, invisible practices (Doeringer *et al.*, 1987), or routinised behaviours.

4) *Understand external linkages*: Move away from a view of industrial districts and other successful regions as 'self-contained' and acknowledge the importance of external linkages, network relations, and non-market connections as one of the determining factors to economic success (cf. Hassink, 1997).

5) *Incorporate the new conceptual insight into comprehensive empirical studies*. Regardless of the past and ongoing conceptual refinement and increasing insight into fundamental issues in technological change at the regional scale, focusing on both the 'hard' and the 'soft' aspects, a lot of empirical research tends to highlight lists of actors involved and engages in 'storytelling' about developments in specific localities. Empirical research still often lacks the ability to dig into the specifics in the empirical cases of the hard-to-measure themes, such as culturally loaded industrial practices, trust-building, coalition building or control relations.

References

Amin, A. and Thrift, N. (1992), 'Neo-Marshallian Nodes in Global Networks', *International Journal of Urban and Regional Research*, vol. 16, pp. 571-587.

Amin, A. and Thrift, N. (1993), 'Globalization, Institutional Thickness and Local Prospects', *Revue d'Économie Régionale et Urbaine*, no. 3, pp. 405-427.

Autio, E. (1998), 'Evaluation of RTD in Regional Systems of Innovation', *European Planning Studies*, vol. 6, pp. 131-140.

Becattini, G. and Rullani, E. (1996), 'Local Systems and Global Connections: The Role of Knowledge', in F. Cossentino, F. Pyke and W. Sengenberger (eds), *Local and Regional Response to Global Pressure: The Case of Italy and Its Industrial Districts*, International Institute for Labour Studies, Geneva, pp. 159-174.

Bowonder, B. (1998), 'Industrialization and Economic Growth of India: Interactions of Indigenous and Foreign Technology', *International Journal of Technology Management*, vol. 15, pp. 622-645.

Braczyk, H.-J., Cooke, P. and Heidenreich, M. (eds) (1998), *Regional Innovation Systems*, UCL Press, London.

Camagni, R. (1995), 'The Concept of *Innovative Milieu* and Its Relevance for Public Policies in European Lagging Regions', *Papers in Regional Science*, vol. 74, pp. 317-340.

Chabbal, R. (1995), 'Characteristics of Innovation Policies, Namely for SMEs', *STI Review*, vol. 16, pp. 103-140.

Chesnais, F. (1993), 'The French National System of Innovation', in R.R. Nelson (ed.), *National Innovation Systems: A Comparative Analysis*, Oxford University Press, New York, pp. 192-229.

Choo, C.W. (1996), 'The Knowing Organization: How Organizations Use Information to Construct Meaning, Create Knowledge and Make Decisions', *International Journal of Information Management*, vol. 16, pp. 329-340.

Cohen, W.M. and Levinthal, D.A. (1990), 'Absorptive Capacity: A New Perspective on Learning and Innovation', *Administrative Science Quarterly*, vol. 35, pp. 128-152.

Conti, S. (1993), 'The Network Perspective in Geography: Towards a Model', *Geografiska Annaler B*, vol. 75, pp. 115-130.

Cooke, P., Gomez Uranga., M. and Etxebarria, G. (1997), 'Regional Innovation Systems: Institutional and Organisational Dimensions', *Research Policy*, vol. 26, pp. 475-491.

De Meyer, A. (1993), 'Management of an International Network of Industrial R&D Laboratories', *R&D Management*, vol. 23, pp. 109-120.

Dicken, P. (1998), *Global Shift*, third edition, Guilford, New York.

Doeringer, P.B., Terkla, D.G. and Topakian, G.C. (1987), *Invisible Factors in Local Economic Development*, Oxford University Press, Oxford.

Ergas, H. (1987), 'Does Technology Policy Matter?', in B.R. Guile and H. Brooks (eds), *Technology and Global Industry: Companies and Nations in the World Economy*, National Academy Press, Washington, pp. 191-245.

Eskelinen, H. and Kautonen, M. (1997), 'In the Shadow of the Dominant Cluster - The Case of Furniture Industry in Finland', in H. Eskelinen (ed.), *Regional Specialisation and Local Environment - Learning and Competitiveness*, NordREFO, Stockholm, pp. 171-192.

Estimé, M.-F., Drilhon, G and Julien, P.-A. (1993), *Small and Medium-sized Enterprises: Technology and Competitiveness*, Organisation for Economic Co-operation and Development, Paris.

Feldman, M.P. and Florida, R. (1994), 'The Geographic Sources of Innovation: Technological Infrastructure and Product Innovation in the United States', *Annals of the Association of American Geographers*, vol. 84, pp. 210-229.

Gertler, M.S. (1995), '"Being There": Proximity, Organization, and Culture in the Development and Adoption of Advanced Manufacturing Technologies', *Economic Geography*, vol. 71, pp. 1-26.

Gertler, M.S. (1996), 'Worlds Apart: The Changing Market Geography of the German Machinery Industry', *Small Business Economics*, vol. 8, pp. 87-106.

Gibb, A.A. (1993), 'Key Factors in the Design of Policy Support for the Small and Medium Enterprise (SME) Development Process: An Overview', *Entrepreneurship and Regional Development*, vol. 5, pp. 1-24.

Glasmeier, A.K., Fuellhart, K., Feller, I. and Mark, M.M. (1998), 'The Relevance of Firm-Learning Theories to the Design and Evaluation of Manufacturing

Modernization Programs', *Economic Development Quarterly*, vol. 12, pp. 107-124.

Grabher, G. (1994), 'The Disembedded Regional Economy: The Transformation of East German Industrial Complexes into Western Enclaves', in A. Amin and N. Thrift (eds), *Globalization, Institutions, and Regional Development in Europe*, Oxford University Press, Oxford, pp. 177-195.

Gu, S. (1996), 'The Emergence of New Technology Enterprises in China: A Study of Endogenous Capability Building via Restructuring', *Journal of Development Studies*, vol. 32, pp. 475-505.

Hassink, R. (1992), *Regional Innovation Policy: Case-studies from the Ruhr Area, Baden-Württemberg and the North East of England*, Netherlands Geographical Studies, Utrecht.

Hassink, R. (1996), 'Technology Transfer Agencies and Regional Economic Development', *European Planning Studies*, vol. 4, pp. 167-184.

Hassink, R. (1997), 'What Distinguishes "Good" from "Bad" Industrial Agglomerations?', *Erdkunde*, vol. 51, pp. 2-11.

Humphrey, J. and Schmitz, H. (1996), 'The Triple C Approach to Local Industrial Policy', *World Development*, vol. 24, pp. 1859-1877.

Humphrey, J. and Schmitz, H. (1998), 'Trust and Interfirm Relations in Developing and Transition Economies', *Journal of Development Studies*, vol. 34, 4, pp. 32-61.

Imai, K. and Baba, Y. (1991), 'Systemic Innovation and Cross-Border Networks: Transcending Markets and Hierarchies to Create a New Techno-Economic System', in *Technology and Productivity: The Challenge for Economic Policy*, OECD, Paris, pp. 389-405.

Johannisson, B. (1993), 'Designing Supportive Contexts for Emerging Enterprises', in C. Karlsson, B. Johannisson and D. Storey (eds), *Small Business Dynamics*, Routledge, London, pp. 117-142.

Johnson, B. (1992), 'Institutional Learning', in B.-Å. Lundvall (ed.), *National Systems of Innovation: Towards a Theory of Innovation and Interactive Learning*, Pinter, London, pp. 23-44.

Jones-Evans, D. and Klofsten, M. (1997), 'Universities and Local Economic Development: The Case of Linköping', *European Planning Studies*, vol. 5, pp. 77-93.

Justman, M. and Teubal, M. (1995), 'Technological Infrastructure Policy (TIP): Creating Capabilities and Building Markets', *Research Policy*, vol. 24, pp. 259-281.

Lambooy, J. and Moulaert, F. (1996), 'The Economic Organization of Cities: An Institutionalist Perspective', *International Journal of Urban and Regional Research*, vol. 20, pp. 217-237.

Lütz, S. (1997), 'Learning through Intermediaries: The Case of Inter-Firm Research Collaborations', in M. Ebers (ed.), *The Formation of Inter-Organizational Networks*, Oxford University Press, Oxford, pp. 220-237.

MacLeod, G. (1997), '"Institutional Thickness" and Industrial Governance in Lowland Scotland', *Area*, vol. 29, pp. 299-311.

Maillat, D. (1998), 'Interactions between Urban Systems and Localized Productive Systems: An Approach to Endogenous Regional Development in Terms of Innovative Milieu', *European Planning Studies*, vol. 6, pp. 117-129.

Malecki, E.J. and Veldhoen, M.E. (1993) 'Network Activities, Information and Competitiveness in Small Firms', *Geografiska Annaler B*, vol. 75, pp. 131-147.

March Chorda, I. (1995), 'Technopolitan Strategies: At the Edge of an Innovation-Driven Territorial Approach', *International Journal of Technology Management*, vol. 10, pp. 894-906.

Maskell, P. (1997), 'Localised Low-Tech Learning in the Furniture Industry', in H. Eskelinen (ed.), *Regional Specialisation and Local Environment - Learning and Competitiveness*, NordREFO, Stockholm, pp. 145-170.

Maskell, P., Eskelinen, H., Hannibalsson, I, Malmberg, A. and Vatne, E. (1998), *Competitiveness, Localised Learning and Regional Development*. Routledge, London.

Moulaert, F., Scott, A.J. and Farcy, H. (1997), 'Producer Services and the Formation of Urban Space', in F. Moulaert and A.J. Scott (eds), *Cities, Enterprises and Society on the Eve of the 21st Century*, Pinter, London, pp. 97-112.

Myllyntaus, T. (1990), 'The Finnish Model of Technology Transfer', *Economic Development and Cultural Change*, vol. 38, pp. 625-643.

OECD (1995), *New Dimensions of Market Access in a Globalising World*, Organisation for Economic Co-operation and Development, Paris, pp. 149-168.

Oinas, P. and Packalén, A. (1998), 'Four Types of Strategic Inter-Firm Networks – An Enrichment of Research on Regional Development. *Terra*, vol. 110, 1, pp. 69-77 (*in Finnish*).

Ostry, S. (1990), *Governments and Corporations in a Shrinking World: Trade and Innovation Policies in the United States, Europe and Japan*, Council on Foreign Relations, New York.

Park, S.O. (1997), 'Networks and Embeddedness in the Dynamic Types of New Industrial Districts', *Progress in Human Geography*, vol. 20, pp. 476-493.

Park, S.O. and Markusen, A.R. (1995), 'Generalizing New Industrial Districts: A Theoretical Agenda and an Application from a Non-Western Country', *Environment and Planning A*, vol. 27, pp. 81-104.

Pavitt, K. (1997), 'Transforming Centrally Planned Systems of Science and Technology: The Problem of Obsolete Competencies', in D.A. Dyker (ed.), *The Technology of Transition: Science and Technology Policies for Transition Countries*, Central European University Press, Budapest, pp. 43-60.

Radošević, S. (1997), 'Technology Transfer in Global Competition: The Case of Economies in Transition', in D.A. Dyker (ed.), *The Technology of Transition: Science and Technology Policies for Transition Countries*, Central European University Press, Budapest, pp. 126-158.

Ruigrok, W. and van Tulder, R. (1995) *The Logic of International Restructuring*, Routledge, London.

Sako, M. (1992), *Prices, Quality and Trust: Inter-firm Relations in Britain and Japan*, Cambridge University Press, Cambridge.

Schein, E.H. (1996), 'Three Cultures of Management: The Key to Organizational Learning', *Sloan Management Review*, vol. 38, 1, pp. 9-20.

Sternberg, R. (1996), 'Technology Policies and the Growth of Regions: Evidence from Four Countries', *Small Business Economics*, vol. 8, pp. 75-86.

Vaessen, P. and Wever, E. (1993), 'Spatial Responsiveness of Small Firms', *Tijdschrift voor Economische en Sociale Geografie*, vol. 84, pp. 119-131.

Vatne, E. (1995), 'Local Resource Mobilization and Internationalization Strategies in Small and Medium Sized Enterprises', *Environment and Planning A*, vol. 27, pp. 63-80.

von Hippel, E. (1994) '"Sticky Information" and the Locus of Problem Solving: Implications for Innovation', *Management Science*, vol. 40, pp. 429-439.

Walker, W. (1993), 'National Innovation Systems: Britain', in R.R. Nelson (ed.), *National Innovation Systems: A Comparative Analysis*, Oxford University Press, New York, pp. 158-191.

Zeira, Y., Newburry, W. and Yeheskel, O. (1997), 'Factors Affecting the Effectiveness of Equity International Joint Ventures (EIJVs) in Hungary', *Management International Review*, vol. 37, pp. 259-279.

Ziegler, J.N. (1992), 'Cross-National Comparisons', in J.A. Alic, L.M. Branscomb, H. Brooks, A.B. Carter and G.L. Epstein, *Beyond Spinoff: Military and Commercial Technologies in a Changing World*, Harvard Business School Press, Boston, pp. 209-247.

Index